D0934238

Praise for
A Scientific Revolution

"The legendary Johns Hopkins Hospital has finally found its bards; their names are Ralph Hruban and Will Linder . . . My prescription, then, is to turn the page and begin reading the sterling essays on the medical disrupters that follow. I am confident that all who do, will savor every chapter."—**From the Foreword, Howard Markel, MD, PhD, George E. Wantz, MD Distinguished Professor of the History of Medicine and Director, Center for the History of Medicine, the University of Michigan Medical School**

"Ralph Hruban and Will Linder's history of medical advances through the lives of medical pioneers is a fascinating history that should be read by every American who enjoys the benefits of modern medicine. This is biological history at its best."—**Robert Dallek, Pulitzer Prize nominated historian**

"*A Scientific Revolution* puts the lives of ten remarkable women and men under the microscope to provide a compelling perspective of more than a century of progress in American medicine. Its strength lies in the honesty of its storytelling. Hruban and Linder bring us into a world where institutional politics, sexism, and racism created severe barriers not only to health equity but to scientific accomplishment. We learn of spectacular advances in medical research and clinical care occurring at a time when personalities often clashed, women were denied equal opportunity, and segregation extended to separate rooms for bodies after death. It's this honesty that makes the book not just a fascinating look into the past but a valuable lens for looking ahead."—**Jon LaPook, MD, Chief Medical Correspondent, CBS News, Mebane Professor of Gastroenterology, Professor of Medicine, NYU Grossman School of Medicine and NYU Langone Health**

"The names Edison, Einstein and Pasteur stand out as inventors, trailblazers and visionaries who changed our world. But in the field of medicine, there are other names we should know. Over a century ago, these ten men and women pioneered how doctors were trained, developed techniques for modern surgery, addressed hygiene issues, and more, all while making great personal sacrifices and enduring hardship. Together, their contributions were transformative. These engaging profiles by Ralph Hruban and Will Linder show how the collective impact of these four women and six men laid the foundation for today's rigorous standards for patient care and clinical research."—**David Louie, Emmy Award-winning Business & Technology Reporter, ABC7 News Bay Area, and past national chairman of the National Academy of Television Arts & Sciences**

"An enthralling and honest history of the first century of scientific medicine in the form of penetrating portraits of ten pioneers. Overcoming obstacles that included addiction, deafness, rampant sexism, vicious racism, and hard-shelled tradition, the ten made possible the medicine of today. Their courage and resilience are a bracing example."—**Scott Shane, former *New York Times* reporter, author of *Dismantling Utopia* and *Objective Troy* and winner of the Pulitzer Prize**

"From the late nineteenth to the mid-twentieth century, the Johns Hopkins Medical Institutions pioneered new procedures and treatments that put patients' health, comfort, and safety foremost. Beginning with John Shaw Billings, who witnessed firsthand the horrors of Civil War medicine and who subsequently gave hospitals their first modern design, to Vivien Thomas, a grandson of enslaved people, who tolerated bitter racial discrimination while pioneering new procedures in heart surgery, here are ten compelling portraits of men and women engaged in a great scientific revolution."—**Ric Cottom, historian and host of WYPR's "Your Maryland"**

A SCIENTIFIC REVOLUTION

Detail from bookplate illustrated by Max Brödel for the surgeon J. M. T. Finney. The illustration, a portion of which is used on the cover, depicts either Hippocrates or the Greek god of healing, Asclepius, using the "Rod of Asclepius," a serpent-entwined staff that symbolizes healing and medicine, to ward off death, depicted wielding a scythe. The work beautifully illustrates the power of Brödel's pen-and-ink illustrations. Though Brödel chose a classical style for this decorative piece, his medical illustrations embody the deep commitment to scientific research exemplified by the ten men and women profiled in this book. (Pen and ink, 1912.)

A SCIENTIFIC REVOLUTION

Ten Men and Women Who Reinvented American Medicine

RALPH H. HRUBAN, MD
& WILL LINDER

PEGASUS BOOKS
NEW YORK LONDON

A SCIENTIFIC REVOLUTION

Pegasus Books, Ltd.
148 West 37th Street, 13th Floor
New York, NY 10018

First Pegasus Books cloth edition May 2022

Chapter opening illustrations by David Rini,
copyright © Department of Art as Applied to Medicine,
the Johns Hopkins University

Interior design by Maria Fernandez

Library of Congress Cataloging-in-Publication Data is available.

ISBN: 978-1-63936-147-2

10 9 8 7 6 5 4 3 2

Printed in the United States of America
Distributed by Simon & Schuster
www.pegasusbooks.com

To Jarmila Hruban. From a small village in Southern Bohemia, to a refugee camp in Germany, to the South Side of Chicago—the arc of her life's journey makes her my "eleventh great."

— R. H. H.

To my wife Jan. In the words of Bob Dylan, "Without your love, I'd be nowhere at all."

— W. L.

It is only in the light of the past that we can hope to solve the problems of the present, and the future.

—*Sir William Osler*

Contents

Foreword

The legendary Johns Hopkins Hospital has finally found its bards; their names are Ralph Hruban and Will Linder. In this wonderfully written book, they tell the stories of ten extraordinary men and women in the history of American medicine. All of them were connected in some capacity to the Johns Hopkins Hospital and Medical School. They were all integral actors in the rapidly changing medical landscape over the past century and a half.

The authors begin with the two "revolutionary philanthropists" who put up the money for what became one of the most recognizable medical entities in the world. We learn about Johns Hopkins, a nineteenth-century Quaker merchant from Baltimore, whose $7 million endowment to establish a university (1876), a hospital (1889), and a medical school was then the largest in the world. The other critical patron was Mary Elizabeth Garrett, a railway heiress, suffragist, and practitioner of what became known as "coercive philanthropy." She insisted that her much-needed financial gift to open the medical school in 1893 was contingent upon allowing women students to matriculate "on the same terms as men."

Although the physician William Osler initially objected to Garrett's proposition, we are all beneficiaries of the always diplomatic and strategic pathologist William Henry Welch, who quickly embraced the terms of the gift and took the money. Twenty-three years earlier, in 1870, the University of Michigan Medical School had already begun accepting female students, making it one of the first major American universities to do so. Aside from the women-only schools or the soon-to-be-closed proprietary medical schools that did accept women in the 1890s, however, Johns Hopkins was one of the few major university-based medical schools to educate both men and women doctors at the fin de siècle. But the male professors at Johns Hopkins required a feminist benefactress to make this then-revolutionary change.

Hruban and Linder next introduce John Shaw Billings. Few people today recollect that the organizer of the Library of the Surgeon General's Office, the president of the New York Public Library, and the designer of the original Johns Hopkins Hospital—including its iconic dome towering over Baltimore—was the same man: Billings. In my humble opinion, Billings was one of the most underappreciated medical geniuses of the late nineteenth and early twentieth centuries. The more people who know about his many accomplishments, the better!

The reader is then taken on a rollicking literary tour through the lives of three of the four founding doctors at the hospital: the internist and role model extraordinaire, William Osler; the statesman of American medicine, William Henry Welch; and the world-class surgeon, as well as raging cocaine and morphine addict, William Stewart Halsted. All were great men and great doctors, but it is also important to recall that these masterful physicians were all too human and subject to the same flaws, foibles, and faults as the rest of us.

The authors also discuss the heroic life of the yellow fever hunter Jesse Lazear. A former Johns Hopkins physician, Lazear died in Cuba assisting Walter Reed during his famous 1900 Yellow Fever Board. Entirely susceptible to the virus, Lazear was an experimental subject. He placed a test tube containing buzzing mosquitoes, all of them carrying the scourge, against his skin. The young doctor was bitten and infected as a result. Soon after, he died a grisly, painful, vomit-stained yellow fever death. For more than a century thereafter, all who stride through the old Johns Hopkins Hospital's main corridor pass a weathered bronze plaque noting the work of Jesse Lazear. The plaque's epitaph is meant to be inspiring—even more so for the impressionable interns who take the time to look up his life story:

> With more than the courage and devotion of the soldier, [Dr. Lazear] risked and lost his life to show how a fearful pestilence is communicated and how its ravages may be prevented.

To this day, medical ethicists and historians have deliberated over the coercion and uninformed consent Walter Reed likely misapplied on the Cuban villagers and the army personnel participating in his experiments. To be sure, Reed scientifically proved the mosquito was the vector of yellow fever. But the human lives lost and sickened would never be tolerated today.

The final chapters of *A Scientific Revolution* focus on five medical pioneers whose careers have only recently been appreciated by historians, including the German émigré and medical illustrator Max Brödel. An anatomical artist, he drew the figures for many of the best scientific papers to come out of Johns Hopkins from 1894 to 1941. This coincided with a period in American history that included two world wars against Germany and during which he suffered a great deal of ostracism and harassment because of his nationality.

The authors also tell the stories of two eminent women physicians who faced enormous obstacles because of their gender: pediatrician and cellular pathologist Dorothy Reed Mendenhall (who, among other things, identified what became known as the Reed-Sternberg cell, a sentinel sign of Hodgkin's lymphoma) and pediatric cardiologist Helen Taussig (who helped develop the blue baby operation). The last chapter is devoted to Vivien Thomas, an African American man who, despite having no formal medical training, ran surgical chief Alfred Blalock's laboratory—first at Vanderbilt (1930–1940) and then at Johns Hopkins from 1940 to 1964, often without credit. Thomas stayed on at the hospital for fifteen years after Blalock's death in 1964, serving as director of the Johns Hopkins Surgical Laboratory.

The authors know that many may ask, "Why ten men and women from Johns Hopkins?" To this likely query, they declare, "mea culpa," and even offer their university affiliations as a possible reason. But I think their response is too modest. One hardly needs an aging historian of medicine to declare the obvious. The Johns Hopkins Hospital played an instrumental role in pulling American medicine out of the muck and mire of nineteenth-century humoralism, bloodletting, and industrial-strength toxins posing as therapeutics. And to this day, the Johns Hopkins Hospital remains a beacon of medical research, education, and practice.

As a former intern, junior and senior resident, clinical and postdoctoral fellow, PhD graduate student, alumnus, and always a gushing fan of what used to be charmingly called "the Hopkins," I cannot tell you how thrilled I was to be reacquainted with (and to learn even more about) these remarkable men and women.

My prescription, then, is to turn the page and begin reading the sterling essays on the medical disrupters that follow. I am confident that all who do will savor every chapter.

—Howard Markel, MD, PhD
George E. Wantz, MD Distinguished Professor of the History of Medicine and Director, Center for the History of Medicine
Professor of Pediatrics and Communicable Diseases
Professor of History
Professor of English Literature and Language
Professor of Health Management and Policy (Public Health)
Professor of Psychiatry
The University of Michigan Medical School

Preface

These vignettes of ten extraordinary men and women in the history of American medicine started out as a series of virtual lectures presented by Ralph Hruban, Baxley Professor and director of the Department of Pathology in the Johns Hopkins School of Medicine, during the global COVID-19 pandemic. At the urging of Will Linder, a veteran writer who was in the online audience and was fascinated with these narratives, they have taken on new life as a book. The timing and format of the original lectures were driven by the pandemic, but their content also resonates deeply with this twenty-first-century crisis.

COVID-19 has posed one of the greatest challenges to medical science the modern world has known. It has summoned enormous effort and sacrifice on the part of medical professionals—not just doctors and nurses but every single caregiver and healthcare worker—greater effort and sacrifice than we could have imagined. For these reasons alone, it is vital to understand how we developed the weapons of science, research, and best practice in caring for patients that we are mobilizing in this epic struggle.

This book, though not a comprehensive history of medicine in a pivotal era, tells the stories of ten men and women—all connected with Johns Hopkins, because that is where the revolution in scientific medicine in America started—who set us on our current path toward medicine that is being driven both by rapid advances in technology and a deeper understanding of fundamental genetic and biological processes.

To the obvious question "Why ten men and women from Johns Hopkins?" we plead mea culpa. We both know Hopkins well as students and alumni (and in Ralph Hruban's case as someone who has spent nearly his entire career at the Johns Hopkins Medical Institutions). And, despite all of its imperfections and shortcomings—many of which we discuss candidly in this book—we have a deep respect and affection for its history. Nonetheless, our choice of ten figures connected with Hopkins is less a matter of being boosters than it is a question of time and place. We believe—and we will argue in this book—that the revolution in scientific medicine that took place in America at the end of the nineteenth century and in the first half of the twentieth could not have occurred without the pivotal contributions of the Hopkins men and women profiled here.

In the spirit of full disclosure, we don't profess to be historians, much less medical historians. One of us is a pathologist who studies pancreatic cancer; the other is a writer and editor. But we are passionate about history and especially the history of medicine. We are sharing these compelling stories about ten innovators who changed the course of medicine in the hope that all of us can apply important lessons from their richly dramatic lives to our own complex and challenging times.

Two more notes of caution may be in order since these relatively short portraits are an exercise in biography. It was William Osler, one of our medical pioneers and the first head of the Department of Medicine at Hopkins, who wrote, "What more delightful in literature than biography? And yet, how uncertain and treacherous is the account which any man can give of another's life."[1]

Osler is warning us that telling the story of another human being's life can be terribly biased. Moreover, we caution that biography is, necessarily, incomplete, focusing as it does on the life of a single individual and setting aside or placing in the background the times.

—Ralph H. Hruban, MD
Will Linder, MBA, MLA
Baltimore and Chicago

INTRODUCTION

A Revolution in American Medicine

The opening of the Johns Hopkins University School of Medicine in 1893 fundamentally changed medicine in America. In the years that followed, medicine was transformed from a trade carried on by poorly trained craftsmen to a science practiced by highly educated physicians. While this transformation may seem perfectly natural in retrospect, the challenges were significant.

We have tried in this book to tell the story of the creation and evolution of the Johns Hopkins Hospital and the Johns Hopkins University School of Medicine through the lives of ten extraordinary individuals: Mary Elizabeth Garrett, John Shaw Billings, William Henry Welch, William Osler, William Stewart Halsted, Jesse Lazear, Max Brödel, Dorothy Reed Mendenhall, Helen Taussig, and Vivien Thomas. Each faced enormous challenges, and each had an unimaginable impact. This account is as much about vision and resilience as it is about medicine and science. We can learn much from the way these individuals lived their lives and pursued their goals.

Uncompromising Vision

Because some readers may be unfamiliar with medical history or Johns Hopkins (or possibly with both), a "sneak preview" of our cast of characters may be helpful. With their uncompromising vision, Mary Elizabeth Garrett and John Shaw Billings created Johns Hopkins Medical School and set the standards that would define Johns Hopkins medicine. Garrett's philanthropy rescued the university and allowed the medical school to be built. Despite great resistance from the first president of the university, Daniel Coit Gilman, she insisted that the medical school set high standards and that it admit women on an equal basis as men.

A veteran of the bloody Civil War battles at Gettysburg and Chancellorsville, John Shaw Billings was asked to design the buildings that would comprise the original hospital (color plate 1). He did much more than simply lay out Hopkins Hospital's leading-edge design principles and plans based on the latest hygiene and sanitation knowledge. He also helped establish the philosophy of the medical school—the belief in science applied to medicine—and he handpicked several of the school's founding faculty members. His accomplishments after that, including his instrumental role in the creation of the National Library of Medicine and the New York Public Library, were just as impressive.

Role Models for Generations

William Henry Welch and Sir William Osler were towering figures for future generations of physicians. Welch, the first head of the medical school, was born in a small town in Connecticut. Described by *Time* magazine late in his distinguished career as the "Dean of American Medicine," "Popsy," as he was fondly called by his students, inspired and trained the first generation of physician-scientists in this country. He promulgated a uniquely forward-looking vision of medicine rooted in scientific research and, in so doing, helped Hopkins forever transform medicine in America.

A Canadian by birth and training, William Osler has been called the greatest physician North America has ever produced. He brought students into the patient wards and taught them at the bedside. Osler's textbook, *The Principles and Practice of Medicine*, was first published in 1892, and for decades was the most authoritative and comprehensive guide for practicing physicians and medical students. Going well beyond his achievements as a clinician and educator, Osler also articulated a philosophy that defined what it means to be a physician. Though it does not do justice to the breadth and depth of Osler's thinking about medicine and the role of the physician, his adage that "to serve the art of medicine as it should be served, one must love his fellow man" is as applicable today as it was a hundred years ago.[1]

Sacrifice in the Name of Science

Where there is scientific triumph in medicine, there is sometimes personal tragedy as well. We tell the stories of two dedicated physicians who made unimaginable personal sacrifices to fight human suffering. The renowned surgeon John Cameron has called William Stewart Halsted, the first chief of surgery at Hopkins, "the most innovative and influential surgeon the United States has produced."[2] From the introduction of surgical gloves to the promulgation of careful and safe operating techniques, as well as by introducing residency training for surgeons, Halsted fundamentally transformed surgery. He did so despite carrying a terrible and crushing personal burden of morphine addiction.

Jesse Lazear, a young physician at Hopkins, volunteered to fight a deadly infectious disease that had ravaged the Americas, including the United States, for centuries. In 1899, at the end of the Spanish-American War, he joined the US Army's Yellow Fever Board in Cuba. Despite its devastating toll, yellow fever's cause and method of transmission had remained a mystery. Lazear was sufficiently persuaded by the accumulating but piecemeal evidence that a species of mosquito was the disease's carrier. He tried a bold—some would say reckless—experiment. Lazear allowed himself to be bitten by a mosquito that had previously fed on the blood of a yellow fever patient. He died an agonizing death two weeks later. The plaque paying tribute to Lazear that hangs in the halls of Johns Hopkins Hospital reads: "With more than the courage and devotion of a soldier, he risked and lost his life to show how a fearful pestilence is communicated and how its ravages may be prevented."

Facing Discrimination

Finally, we chronicle the lives of four "outsiders" at Hopkins, an immigrant, two women, and a Black man. Each of these lives was marked by great accomplishment despite blatant discrimination. One of the four, Max Brödel, brought the power of scientific illustration to medicine. Dorothy Reed Mendenhall made groundbreaking discoveries in cancer research. Two of the others, Helen Taussig and Vivien Thomas, played leading roles in the birth of cardiac surgery, in particular the pioneering "blue baby" operation that saved thousands of children born with congenital heart defects.

With no formal science training, Brödel, an immigrant from Leipzig who experienced firsthand the anti-German sentiment so prevalent during World War I, collaborated with Howard Kelly, William Halsted, Harvey Cushing, and many others at Hopkins, advancing their surgical work and teaching with his unparalleled medical illustrations. Called "the greatest anatomical artist since Leonardo da Vinci,"[3] Brödel created illustrations that were critical in promulgating the discoveries made at Hopkins. Despite being an outsider and a non-physician, Brödel's impact on medicine would come to be as great as that of any of the doctors whose work he so dramatically illustrated.

Reed Mendenhall was a member of the School of Medicine's class of 1900. A year later, as a fellow in William Welch's pathology lab, she characterized the cell that causes Hodgkin disease. Her reward? Reed was denied a faculty position at Hopkins because of her sex. Born to wealthy parents in Columbus, Ohio, she defied her family's expectations that she would "marry well" and become an ornament on that city's social scene and went on to study at Smith College and then at Hopkins. Reed endured not only discrimination as a woman but also the loss of two children. Nonetheless, she left a lasting mark as a researcher and as an advocate for mothers and their babies.

The "Mother of Pediatric Cardiology," Taussig was a Phi Beta Kappa graduate of Berkeley and wanted to study public health. The dean of the Harvard School of Public Health told her that, because she was a woman, she could attend classes but would not be awarded a degree. She enrolled at Johns Hopkins Medical School instead. Her clinical skills were so great that, even though she was nearly deaf later in life, she was able to use her fingers to detect a baby's heart abnormality.

The grandson of an enslaved person, Vivien Thomas conducted the critical experiments on dogs that literally charted the path to the blue baby heart operation at Hopkins. He stood at the shoulder of the famous surgeon Alfred Blalock, who had brought Thomas with him to Hopkins, and guided him during the first operation on a patient. Blalock described Thomas's surgery as masterful: "This looks like something the Lord made."[4] Thomas instructed generations of Hopkins residents in surgical technique. Yet, because of his race, Thomas was forced to remove his white laboratory coat before entering the corridors of Hopkins Hospital. Only at the end of his career was Thomas appointed to the medical school faculty and awarded an honorary doctorate by the university.

Along with so many firsts and remarkable achievements, the story of these ten Hopkins figures who transformed the course of medicine in America is one of exceptional foresight—a foresight that once and for all aligned medicine with the forces of science, public health and hygiene, and patient-centered care. That said, the story also has inextricably woven into its fabric the ugly realities of racism, sexism, and a host of other harsh truths. While some of these harsh truths may be understood in the context of their place and their time, this understanding does not absolve past wrongs. Furthermore, as we learn about these ten men and women, we recognize that people, even those who do extraordinarily good things to help others, can hold strong biases, both conscious and unconscious—as we are becoming aware of today. We have taken pains not to gloss over these unpleasant realities so that the lessons of these lives can better prepare us to face a complex and challenging future.

1

MARY ELIZABETH GARRETT

Equity and Excellence

Mary Elizabeth Garrett might seem an unlikely figure to ignite a revolution that transformed the way American physicians are educated and, indeed, what sorts of people become medical doctors in this country. Garrett walked with a limp and was lonely as a child. She never attended college, despite her keen intellect, and she was almost completely absorbed in her father's business affairs from the time she was a teenager. Yet, Garrett was an extraordinary figure, possessed of both the vision to conceive an entirely new future for the education of women and the tenaciousness to make that future a reality.

To understand Garrett's seismic impact on the teaching of medicine in the United States, we need to go back to the late nineteenth century, before the 1893 opening of the Johns Hopkins School of Medicine. In those days, as we noted, medical schools were essentially for-profit trade schools. One pundit quoted in the famous Flexner Report, which looked at the state of US medical education, opined that medical schools were filled with young men "too stupid for the Bar" and "too immoral for the Pulpit."[1] When Harvard's president Charles Eliot proposed in 1869 to hold written examinations for the degree of doctor of medicine, he was quickly rebuffed by members of the school's faculty. The renowned professor of surgery Henry Bigelow declared, "He [Eliot] knew nothing about the quality of Harvard medical students. More than half of them can barely write. Of course, they can't pass written examinations."[2]

Bigelow's objection to written exams tells us a great deal about medical education at the time. At least one hundred medical schools would accept anyone willing to pay. Fewer than 20 percent required a high school diploma.[3] In 1870, a Harvard medical student could fail four of nine courses and still get a degree.[4] Medical schools in those days often offered two-year programs, and the first year's curriculum was repeated in the second so that students who had failing grades after a year of study had another chance to earn a diploma. Even more alarming, most students could graduate without ever touching a patient, if only because it was the rare medical school that was affiliated with a university and rarer still one connected with a hospital. In 1893, only one medical school in the United States—the brand-new Johns Hopkins—required students to have a college degree.[5] Hopkins also required students to be fluent in French and German and, notably, to have a strong background in science, but such requirements were almost unprecedented.

When it opened, largely due to the foresight and generosity of Mary Elizabeth Garrett, the Johns Hopkins School of Medicine created an entirely new path in medical education. High admission standards ensured that the best and the brightest students enrolled. These students routinely visited the wards, so they could see and learn from and about patients.

New scientific discoveries were one of the school's primary goals, and these, in turn, were systematically applied to the treatment of patients. In short, Hopkins rather quickly became "a model of its kind" in the training of physicians and the practice of scientific medicine across the nation and throughout the world.[6] The full unfolding of these advances in science-based medicine still lay in the future at the end of the nineteenth century, but with this perspective on medical education at the time, we can turn our attention to Mary Elizabeth Garrett, the woman who made the Johns Hopkins School of Medicine possible and set it on its historic course.

A Lonely Childhood

Wealth and influence were Mary Elizabeth Garrett's birthright. Yet, like many American success stories, the Garrett family saga had a humble beginning. Mary Elizabeth's paternal grandfather Robert Garrett was a poor immigrant from northern Ireland. In 1790, when Robert was seven, the family sailed to America aboard the brig *Brothers*. Robert's father died on the lengthy voyage, and his remains were buried at sea. (Ships like these, jammed with emigrants, were rightly dubbed "coffin ships.")[7] Despite this tragic start, Robert prospered in his adopted country. He made a fortune hauling goods in Conestoga wagons over the National Road, the first federal highway, which linked the Potomac and Ohio Rivers. The transition from wagons to railroads was swift. By 1858, Mary Elizabeth Garrett's father, John Work Garrett, had been named president of the Baltimore and Ohio Railroad, the nation's first commercial railroad. Through relentless hard work and keen entrepreneurial ability, he soon became the most powerful man in Maryland.[8]

Mary Elizabeth Garrett was born in 1854. If privilege was part of her inheritance, good fortune was not. As she later described it, "when I was about eight months old, a very serious trouble with the bone of the right ankle developed."[9] For some years, Garrett wore a brace, first made of iron and later of whalebone. The brace made her feel unattractive, and she became shy. She wrote, "I was heavy and the lameness made me less active than ordinary children and also more solitary"[10] (figure 1.1).

FIGURE 1.1. *Mary Elizabeth Garrett as a young girl, circa 1865. (Photo by J. H. Dampf & Company, Baltimore.)*

Garrett had a private education, studying mainly at Miss Kummer's School for Girls on Baltimore's desirable Mount Vernon Place and with private tutors when she traveled abroad with her family. She was an intelligent young woman, but as was typical in the nineteenth century, she was thwarted in her desire for a higher education. Mary's two brothers were sent to Princeton, but she was prevented from going to college. "I begged my father that I should be allowed to go to college."[11] He was adamant in his refusal. When Johns Hopkins University opened its doors in 1876, Garrett again tried to pursue a higher education, but the school's president, Daniel Coit Gilman, declined to admit her, allegedly on grounds of caution. Gilman had made up his mind that young women should "not be exposed to the rougher influences which I am sorry to confess are still to be found in colleges and universities where young men resort."[12]

The Friday Night

Garrett joined forces with a group of friends, several of whom were similarly thwarted in their aspirations for an equal education. They were the daughters of prominent Baltimore leaders. All but one of their fathers were trustees of the newly founded Johns Hopkins University. M. Carey Thomas, Mamie Gwinn, Bessie King, Julia Rogers, and Mary Elizabeth Garrett met on the second Friday of each month for serious conversations. They called the group "the Friday Night," or "the Friday Evening" (figure 1.2). Their focus, Garrett's biographer Kathleen Waters Sander writes, was on the "woman question."[13] Could women succeed and contribute to society at the same level as men? According to one Thomas scholar, "The Friday Evening gave [the group] a context in which to think out some of the dilemmas of choice . . . an occasion for these talented women, in their early twenties and living with their families, to meet and talk seriously about life, religion, vocation, and, not incidentally, marriage."[14] Thomas and Garrett would develop their own solution to the marriage dilemma, but not before Thomas had declared, "it is difficult to conceive [of] a woman who really feels her separate life work to give it all up when she marries a man."[15]

FIGURE 1.2. *The Friday Night with Garrett in the center. The others are (clockwise from the upper right) Elizabeth King, Julia Rogers, M. Carey Thomas, and Mamie Gwinn.*

Mary Elizabeth Garrett's brothers, Harry and Robert, were given powerful positions in the Baltimore and Ohio Railroad, but they would ultimately fail to live up to their father's expectations for carrying on his legacy.[16] Mary was relegated, as she put it, to being "Papa's secretary."[17] She often traveled with her father on railroad matters. Though she never attended college, Garrett readily absorbed the details of the family business and learned a great deal about the railway as her father met with his colleagues and contacts, including giants in American business like Andrew Carnegie, J. P. Morgan, and Cornelius Vanderbilt.[18] In his later years, John Garrett may have come to regret his decision to deny his daughter a college education and a career. He was known to have remarked to a close friend, "I have often wished in these last few years that Mary was a boy. I know she could carry on my work, after I am gone."[19]

Equally important were the lessons Garrett learned from her father about how to use great wealth and influence to shape the course of events. She would later draw liberally on these lessons to advance her goals.[20] As her father's health failed, Garrett took over more of his affairs. John Garrett died in 1884. At the age of thirty, Mary Elizabeth Garrett inherited one of the largest fortunes of the day. Though he had not made it known to his daughter before his death, John Garrett had bequeathed Mary one-third of his fortune, which included $2 million and three lavish estates.[21]

Garrett's use of her huge inheritance was remarkable in its scope and impact. From the start, she set out to right the wrongs of unequal access and unfair treatment that society had imposed on women. Her gender had denied her a college degree. Step by step, she built a path for women's education. In one of her first philanthropic acts, Garrett helped to fund the establishment of the Bryn Mawr School for Girls, an institution in Baltimore founded by the women of the Friday Night, whose purpose was to prepare girls for the rigors of a postsecondary curriculum. She wrote that "the prescribed course will be so arranged as to include the highest requirements for entrance made by any college."[22] Bryn Mawr was, in the words of the former headmistress who spoke at the school's seventy-fifth anniversary celebration, "the only School for girls in those days from the Atlantic to the Pacific where every girl had to take the college preparatory course and every graduate had to pass the college entrance examinations."[23]

Garrett and the other women of the Friday Night fought long and hard to create a school to prepare girls for a college education. With only rare exceptions, the men who controlled the levers of power and decision-making in Garrett's day simply saw no purpose for such an institution. As we noted, Daniel Gilman had been named the first president of Johns Hopkins University a decade earlier. Garrett wrote to the Friday Night that Gilman felt the whole idea of women's higher education was pointless because "women were so different from men, that they had not the same need of education as men and for them college education was oftener a liability than an asset."[24] A Baltimore physician of that era advised girls to "avoid the school [Bryn Mawr] because climbing the stairs would affect future childbearing."[25] In spite of this profound sexism, the school grew in size and offered a first-rate college preparatory program for young women.

FIGURE 1.3. *Gymnasium at the Bryn Mawr School for Girls. Although she herself had felt "lame," Garrett insisted that the new school emphasize physical education.*

Garrett, we recall, had spent much of her childhood in a leg brace, mostly alone and isolated from her able-bodied peers. Nonetheless, the Bryn Mawr School placed particular emphasis on physical education. In fact, at Garrett's direction, it had one of the best-equipped gymnasiums in the United States. She invited Dr. Dudley Allen Sargent, founder of the Sargent School of Physical Culture, to design Bryn Mawr's gymnasium.[26] Garrett herself was reported to have taken a leading role in selecting the apparatus[27] (figure 1.3). Garrett succeeded in creating a school with high standards that prepared young women for college, but her work to advance women's educational opportunities was just beginning. To fully appreciate her role in the vanguard of equal education for women, we need to return to the story of Johns Hopkins and the creation of the medical institutions that bear his name.

A Medical School Open to Women: Dreams and Realities

Johns Hopkins, a close friend of Mary Elizabeth Garrett's father, was a wealthy Baltimore merchant, banker, and investor with strong civic and philanthropic inclinations. Toward the end of his life and shortly after the Civil War, he met with George Peabody, the American financier and founder of the Peabody Institute (the first major arts and intellectual center in an American city), which is now part of Johns Hopkins University, at John Work Garrett's home. Following Peabody's advice to use his fortune to advance higher education, Johns Hopkins drafted his will and appointed a board of trustees to carry out his wishes. Johns Hopkins died on Christmas Eve in 1873, leaving his bequest of $7 million to found a university, a medical school, and a hospital. At the time, his gift was the largest philanthropic donation in US history.

Unusual for its era, Johns Hopkins's bequest tied together the three institutions he intended to establish. His instructions to the trustees declared, "You will bear constantly in mind that it is my wish and purpose that the institution [hospital] should ultimately form a part of the medical school of that university for which I have made ample provision by my will."[28] The first of these three interconnected institutions, the university, opened on October 3, 1876.

Construction of the Johns Hopkins Hospital began in 1877 on the site of the old Maryland Hospital for the Insane, but the university's board of trustees, composed mostly of frugal Quakers, was in no rush, and the hospital was not completed until 1889. Johns Hopkins had, for better or worse, tied his vision for the three institutions to the shares of the Baltimore and Ohio Railroad (B&O) he had donated in his bequest. His will required that the university not sell any of the shares it had received. By 1888, the faltering price of B&O shares was threatening the university with a severe financial crisis and jeopardizing the now-deceased Johns Hopkins's duly recorded intention to open a medical school together with a university and a hospital. The final blow to the institution's finances was B&O's eroding stock dividend. The railroad, which had previously paid out a healthy 8–10 percent of its income, lowered the dividend in the late 1880s and then halted it altogether, reducing the university's annual operating budget by nearly 75 percent.[29] "The fiasco to the Baltimore and Ohio Railroad system," wrote the *Chicago Journal*, "threatened to paralyze the university."[30]

President Gilman became desperate. The hospital was about to open in 1889, but the university had only $67,480 left to fund the School of Medicine.[31] Gilman declared that an additional $100,000 was needed to open the school. The five women of the Friday Night saw an opportunity to advance their agenda and to break down another gender barrier in education. They formed the Women's Medical School Fund Committee. A December 31, 1888, letter from Thomas to Garrett revealed the committee planned to raise $100,000 and donate it to the School of Medicine with the stipulation that women be admitted to the medical school on the same terms as men (figure 1.4). "Mr. Gilman is now totally opposed to the scheme," Garrett's close Friday Night friend wrote her, "much more opposed than I realised . . . Dr. Hurd [the first superintendent of the hospital] and Dr. Osler [are] vehemently in favour of it."[32] In another letter, Thomas reported, "many of the trustees, and Gilman above all, seemed to prefer not to open the school at all if it meant that women were to be admitted."[33]

On October 28, 1890, the Women's Medical School Fund Committee, by now augmented with participants from other parts of the country, presented the university its offer to donate $100,000. The group had in fact raised only about half the sum, but Garrett committed the additional $47,787 needed to meet Gilman's $100,000 demand. The committee pledged the donation under the condition that women would be admitted to the Johns Hopkins School of Medicine. The trustees accepted the offer but declared that the donation was merely the "foundation" for a required $500,000.[34] Suddenly, the bar confronting Garrett and the Women's Fund Committee had been raised fivefold!

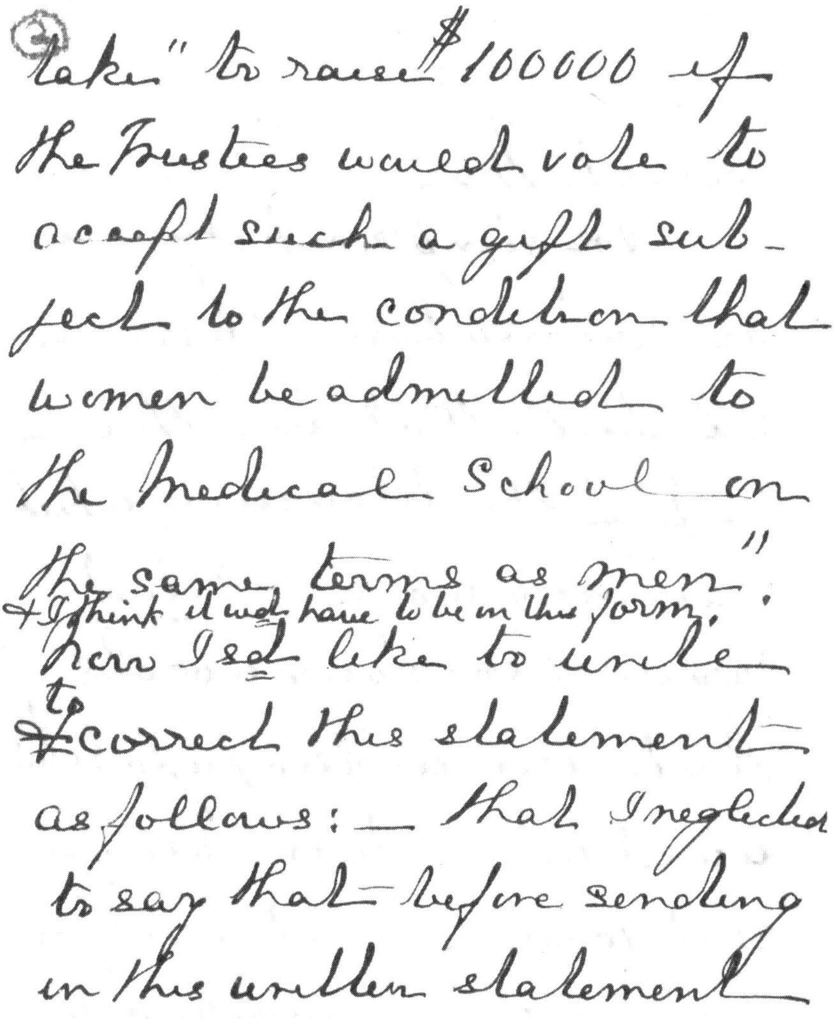

take" to raise $100000 if the Trustees would vote to accept such a gift subject to the condition that women be admitted to the Medical School on the same terms as men". + I think it wd have to be in this form. Now I sd like to write to correct this statement as follows:— that I neglected to say that before sending in this written statement

FIGURE 1.4. *Excerpt of a letter from M. Carey Thomas to Garrett in which Thomas outlines their "scheme" for persuasive philanthropy.*

At this point in our narrative, it is worth considering the almost insurmountable hurdles women faced in gaining admission to medical schools. The prevailing view of a male-dominated medical profession in the nineteenth century was that women were too "frivolous and delicate to handle full-strength medical education, with its gory emphasis on human anatomy and disease."[35] The American Medical Association, the alleged guardian of the integrity and well-being of the profession, was founded in 1846 but refused to admit female physicians as members until 1915.[36] Sexism and misogyny needed no cover or subterfuge in this era. A professor at Harvard Medical School, Edward H. Clarke, claimed that if women were admitted to medical schools, they would inevitably develop "monstrous brains and puny bodies; abnormally active cerebration; and abnormally weak digestion; flowing thought and constipated bowels."[37]

A few exceptional women managed to beat the odds and make it into medical school. (There were nineteen women's medical colleges by 1900, but women who received diplomas from these programs were often denied clinical training after they graduated.)[38] The trials and eventual triumph of Elizabeth Blackwell have been recounted in several well-written books and articles. Suffice it to say that Blackwell was rejected by twenty-nine medical schools before she gained admission to Geneva Medical College, a small institution in upstate New York, in 1847. She was admitted because, and only because, the dean of the school, feeling pressure from an eminent physician who supported Blackwell's admission, decided to abdicate his responsibility and put her case before the all-male student body. Confident that the young men would quickly reject Blackwell, the dean hadn't bargained on the students' sense of humor. Treating it as a lark, the students voted as one to admit Blackwell. She graduated two years later at the top of her class and went on to found the New York Infirmary for Women and Children.[39] In addition to caring for the city's poor, the infirmary employed an all-female physician staff and trained female medical and nursing students.

A Gift with Six Strings Attached

Mary Elizabeth Garrett's response to the Hopkins trustees' escalating financial demands was characteristic—and revealing. She agreed to donate an additional $100,000 of her own money if the trustees could raise the balance, putting the ball squarely back in their court. Gilman and the trustees failed to come up with the amount needed. In November 1892, Garrett's Friday Night ally Thomas sent her a note saying the "outlook is infinitely worse than I had thought."[40] Undaunted, Garrett took matters into her own hands and declared her intention to contribute an additional $306,977, an extraordinary sum for the time. Shrewd as she was forceful, Garrett took the measure of the changeable trustees and offered a sum that would satisfy their financial demand—but not one penny more. She further specified that the money would be paid in six equal installments of $50,000 each beginning on January 1, 1894, the year *after* the medical school's scheduled opening, with a final installment of $6,977 to be paid on January 1, 1899.

With Garrett's commitment to advancing women's equality in education, she could have simply donated the money. Instead, she went beyond philanthropy, practicing what her principal biographer has called "coercive philanthropy," demanding in return for her money a large measure of control over how it would be spent.[41] Garrett outlined to the trustees in her letters of December 22, 1892, and February 15, 1893, unalterable conditions for her gift. In addition to requiring that Hopkins admit women and men to the medical school with "no distinction," Garrett's demands, taken from her correspondence here, stipulated that:

- A building must be constructed to honor the role of women in the founding of the school—the Women's Fund Memorial Building.
- The Johns Hopkins Medical School, like the great European universities, "shall be exclusively a Graduate School . . . and shall form an integral part of the Johns Hopkins University . . . [and] shall provide a four years' course to the degree of Doctor of Medicine."
- The terms of the gift "shall be printed each year in . . . calendars . . . announcing the courses of the Medical School."
- Entering students shall "have a good reading . . . knowledge of French, German," and shall have studied physics, chemistry and biology.

- The school must "leave undiminished" its high "standard of admission."[42]

Garrett also made it clear that Johns Hopkins would have to return her gift if "the University shall discontinue a Medical School devoted to the education of both men and women," or if "women studying in the Medical School do not enjoy all its advantages on the same terms as men."[43] As if this weren't enough to ensure her goals were realized, Garrett also insisted that the new medical school open by the autumn of 1893. This last demand was intended to guard against the trustees' demonstrated tendency to procrastinate.

The response to Garrett's forceful terms was disheartening but not surprising, given the prevailing biases against women's education. Virtually every day, a university trustee, a faculty member, or President Gilman himself visited Garrett's home in the Mount Vernon neighborhood of Baltimore, pleading for concessions. Throughout the struggle, Thomas, Garrett's steadfast supporter, offered her Friday Night colleague unwavering encouragement and occasional tactical advice. At one point, she wrote Garrett, "to yield now will mean no [to admitting women] forever . . . they must think you a fool to believe you, or indeed we, can be hood winked in that way . . . even being a woman you have the whip hand for once + can afford to be absolutely immoveable . . . we are absolutely in the right, magnificently in the right."[44] At another, Thomas apprised Garrett that that the trustees fought "in the dark with treachery and false reasons . . . a tangle of hatred, malice, detraction that beggars description."[45] Fighting with, among others, the very man who denied her a college education, Garrett refused to budge. And in her own tenacious way, she gave as good as she got, ending a January 1893 letter to Gilman and the board of trustees by declaring, with more than a bit of irony, "I regret that my letter did not seem sufficiently distinct upon these points."[46] In the end, financial realities forced Gilman to give in to Garrett's conditions, clearing the path for Hopkins to forever transform medical education in the United States.

Some feared that the high academic standards Garrett insisted on would scare applicants away. As William Welch, the first dean of the medical school, wrote, "It is one thing to build an educational castle in the air at your library table, and another to face its actual appearance."[47] William Osler, founding chair of the Department of Medicine, commented wryly to Welch, "It is lucky that we get in as professors; we never could enter as students."[48]

Garrett's demands did more than establish minimum criteria for admission. The standards she set attracted the very best future physicians to Hopkins. If you were a smart, talented, and driven young man or woman who aspired to a career in medicine, would you want to enroll in a medical school where other students had only a high school education? Or would you rather receive your training at a medical school that had the highest standards and where you would be exposed to accomplished peers? Garrett fundamentally made Johns Hopkins Medical School the extraordinary institution that it is today (figure 1.5).

FIGURE 1.5. *The Johns Hopkins School of Medicine's graduating class of 1898. Unlike other medical schools at the time, Hopkins admitted women on an equal basis with men.*

Bryn Mawr College and the Women's Suffrage Movement

As a result of Mary Elizabeth Garrett's grit and perseverance, three women were admitted to the first entering class at Johns Hopkins Medical School in 1893. That milestone and her earlier establishment of the Bryn Mawr School for Girls in Baltimore were part of her vision and lifelong commitment to advancing educational equity for women. In addition to these twin achievements, Garrett devoted her formidable talents and considerable wealth to Bryn Mawr College. The college, located near Philadelphia, had opened in 1885 and was dedicated to affording women a more rigorous undergraduate education than was available to them at the time. As she had done at the Bryn Mawr School for Girls and at Hopkins, Garrett based her philanthropic vision for the college on the dual pillars of high academic standards and equal opportunities for women.

In 1893, the year that the Johns Hopkins Medical School opened, Garrett offered to donate $10,000 annually to Bryn Mawr College if the trustees elected her close friend and fellow Friday Nighter Thomas to the institution's presidency.[49] The trustees agreed, and, over her lifetime, Garrett contributed nearly $450,000 to Bryn Mawr College.[50] Garrett's generosity to Bryn Mawr College filled the gap between her earlier philanthropy supporting women's preparation for college and her commitment to furthering women's educational opportunities at the graduate and professional level.

As the years went on, Garrett became more closely connected with her lifelong friend M. Carey Thomas. Mamie Gwinn, who had been Thomas's intimate friend and housemate for nearly twenty years at Bryn Mawr College, married Alfred Hodder, a faculty member, in June 1904. The marriage was a blow to the remaining members of the Friday Night, who had all pledged that they would never marry and "degrade" themselves by subjugating themselves to the bondage and lack of freedom that having a husband implied.[51] Thomas and Gwinn never spoke to each other again after Gwinn's marriage. In 1904, Garrett began to winter with Thomas at Bryn Mawr and eventually moved in with her in what was then called a "Boston marriage." The term referred to friendships in which a pair of women lived together. According to one of Thomas's biographers, the arrangement was not at all unusual in the mid-nineteenth century and did not carry with it connotations of sexual behavior.[52]

A year before joining Thomas at Bryn Mawr, Garrett had taken the lead in creating perhaps the most famous painting in American medicine, commissioning John Singer Sargent's portrait *The Four Doctors.*[53] The painting depicts William Welch (Hopkins's first dean and chief of pathology), William Osler (chief of medicine), William Stewart Halsted (chief of surgery) and Howard Atwood Kelly (first chair of gynecology and obstetrics). All were members of the Johns Hopkins School of Medicine's founding faculty, and all were personal acquaintances of Garrett (color plate 2).

The four physicians are dressed not in their clinical attire but in their academic robes—an allusion to their learnedness and academic credentials, fitting for a school with rigorous standards that was charting a new scientific path in medical education. Completed in 1906, the portrait debuted to much critical acclaim at London's Royal Academy. It was then shipped to Baltimore, where it was presented by Garrett to the university.

By this time, Garrett was deeply committed to the women's suffrage movement. As early as 1902, Garrett underwrote the cost of striking a medallion commemorating Susan B. Anthony, a leading advocate for women's rights, and presented it to Bryn Mawr College. At her urging and that of Thomas, the National American Woman Suffrage Association convened in Baltimore in 1906. Garrett entertained Anthony and members of the association at her Mount Vernon mansion in a series of events intended to honor the suffragist leader. Garrett even persuaded medical school dean William Welch to march through Baltimore's streets with the suffragists.

"A Woman Who Served Her Day and Generation Well"

Mary Elizabeth Garrett died of leukemia at Bryn Mawr College on April 3, 1915.[54] She was sixty-one. Garrett is buried next to her father and close to the grave of Johns Hopkins in Baltimore's Green Mount Cemetery, not far from Hopkins Hospital (figure 1.6). Carved on her tombstone are these words:

> A woman of quiet
> realized enthusiasms
> She served her day and
> generation well and will
> be long remembered by
> those for whom she
> laboured

FIGURE 1.6. *Coauthor Ralph Hruban visiting the gravesite of Mary Elizabeth Garrett in Baltimore's Green Mount Cemetery.*

We should long recall Mary Elizabeth Garrett and her contributions to the pioneering medical school that many have argued should in all justice have borne her name. We honor her for her tireless struggle to create institutions based on academic rigor and fairness to women. When we think of Garrett, it is almost impossible not to think of her unbending principles of equity and excellence, which enriched and transformed not only the Johns Hopkins School of Medicine but also the standard of medical education for generations to come (color plate 3).

2

JOHN SHAW BILLINGS

Building Medicine for a New Century

If you were to walk into the domed rotunda of Hopkins Hospital (color plate 1), you would see a dignified painting of founder Johns Hopkins on your left, but if you were to next turn your gaze to the right, you would be looking at another formidable full-length portrait—this one of John Shaw Billings (color plate 4). And, in fact, the building as we know it today is named the Billings Administration Building in his honor.

Mary Elizabeth Garrett set the stage for the new Johns Hopkins School of Medicine, but it was John Shaw Billings who planned a hospital that incorporated the latest patient care and hygiene principles and set the medical school on its historic path toward distinction in research and physician education. Carleton Chapman's lucid biography of Billings is titled *Order Out of Chaos*. This phrase aptly describes Billings's career. Billings had a powerful, seemingly innate, talent for taking sprawling, previously unmanageable problems and bringing them under the rule of reason and predictability. Yet even more striking than his uncanny ability to organize and direct huge undertakings was his extraordinary capacity to conceive and implement plans far beyond the horizons of his era. Billings was a visionary, but that rare visionary who got things done. This unique combination of talents was abundantly evident in the design of the Hopkins medical institutions and in the collection and management of information, two of the arenas in which Billings left an indelible legacy.

A Childhood Spent Reading

Who was this often overlooked and unheralded figure, and how did he come to make so many pivotal contributions to Johns Hopkins and, indeed, the history of scientific medicine? John Shaw Billings was born on April 12, 1838, near Allensville, Indiana, a small town in the southern part of the state.[1] According to an early biographer, Billings "loved books, and his pleasure and recreation consisted in reading them."[2] Billings later recalled that by the age of eight he had read every verse of the Bible, *Robinson Crusoe*, *The Deer Slayer*, and several other classics.[3] As a precocious fourteen-year-old, he entered college at Miami University in Ohio.

Apparently, a higher education for their son was not always in the Billings family's plans. Another biographer suggested Billings had to strike a deal with his father, a financially pressed businessman and itinerant farmer: In exchange for the funds to attend college, Billings agreed that his share of the family inheritance should go to his sister.[4] At Miami, Billings's devotion to reading persisted. The college library was open for only a few hours on Saturdays, and it operated solely during the regular school term.

When summer vacation arrived, Billings became known for "burglarizing the library to read undisturbed for hours on end."[5] He graduated second in his class and went on to medical school in 1858 at the Medical College of Ohio in Cincinnati. Like most medical schools at the time, the school offered a two-year program, and both years were essentially the same to make sure that even the weakest students would receive their degrees. Billings described his classroom experience in these terms: "In those days they taught medicine as you teach boys to swim, by throwing them into the water."[6]

Despite the school's deficiencies, Billings managed to publish a paper on the surgical treatment of epilepsy.[7] The paper was remarkable, mostly from a bibliographic perspective. Billings consulted articles in forty-four journals by fifty-one different authors, most of them European. Hardly any of the journals were available to him at libraries in Cincinnati, but he had access to a mentor's large private collection of articles on surgical topics. Decades later, Billings recalled, "there was not in the United States any fairly good library, one in which a student might hope to find a large part of the literature relating to any medical subject."[8] The seeds of Billings's monumental contributions to medical libraries and medical bibliography had been planted. After graduation, as there were no formal residency training programs in the United States, he secured additional training at the Commercial Hospital in Cincinnati, which by 1860 was affiliated in an embryonic and, for the times, innovative teaching relationship with the Medical College of Ohio. Billings was only loosely supervised by faculty at the Medical College and had to shoulder on his own most of the burdens of caring for the hospital's clientele of prostitutes, river boatmen, and others from the lower rungs of society. As was characteristic of this era, wealthier patients preferred to receive medical and surgical care in their homes.[9]

Surgery Under Fire

Around the time Billings completed his training, the growing sectional rivalries and bitter divisions over slavery that had been building for decades finally exploded when the Civil War broke out in 1861. Billings soon volunteered to be a doctor in the Union Army. His performance before the army's Medical Examining Board was the strongest among all the candidates who applied.[10] Legend has it that Billings scored so highly that his examiners, suspicious that a candidate from Indiana had done so well, asked him to take the test again. The second time, it is said, his performance was even better.

With only minimal training, Billings became a military surgeon. He was ordered to Chancellorsville and then to Gettysburg, two of the bloodiest battlefields imaginable. The life-and-death struggles that played out on wartime operating tables were harrowing. At Chancellorsville, Billings set up his field hospital dangerously close to the front lines. He wrote, "I soon found that the wounded who could walk would not stop where I was—it was entirely too close . . . I moved back about 200 yards and began to work there, but soon got an order from the medical director saying that I was still too close."[11] A building used as a temporary hospital was "shelled and burnt" and one of Billings's assistants was killed.[12]

At Gettysburg, Billings established a hospital in a house and barn near Round Top. From there, on July 9, 1863, he wrote his young wife, "the days creep by and I am still trying to produce order out of chaos and to get my wounded patients into something like the state of comfort."[13] In another letter from Gettysburg, Billings reported, "I am utterly exhausted mentally and physically, have been operating night and day and am still hard at work. I have been left here in charge of 700 wounded with no supplies . . . I had my left ear just touched with a ball."[14]

Laying the Foundation for a Life's Work

Billings survived Chancellorsville and Gettysburg, though the battles took such a heavy toll on his body and his spirit that he had to take a leave of absence to recuperate in a hospital (figure 2.1). His experiences as a battlefield surgeon certainly shaped his outlook on war and its human costs, but his other postings in the military were more important in shaping his future work as the designer of Johns Hopkins Hospital, master bibliographer, and proponent of scientific medicine.

FIGURE 2.1. *Billings before (left) and after (right) his heroic service at Chancellorsville and Gettysburg. His mournful gaze and sunken eyes reveal the deep toll that treating the many wounded and dying had taken on him.*

In 1862, the year before the Battle of Gettysburg, Billings was assigned to the army's General Hospital in Philadelphia. In those days, the city was in the vanguard of American science. There Billings discovered the new research tool of microscopy. As with so many other fields his encyclopedic curiosity touched, Billings became an expert in its use. He would eventually publish an article discussing the application of microscopy to diagnosing infections in cattle. More importantly, Billings became ardently committed to the idea that laboratory investigation was the foundation of effective medical treatment—a concept that he would carry forward in his work designing the Johns Hopkins medical institutions.

As the Civil War entered its final phase, Billings joined the Office of the Surgeon General in Washington, DC (figure 2.2). He would remain there for more than thirty years, though at times combining his public health role with key leadership positions in other important medical and governmental institutions. Polymath that he was, Billings was able to harmonize and even leverage these complex and often competing demands on his time.

FIGURE 2.2. *Surgeon General Joseph K. Barnes (standing, center) and his staff in the Office of the Surgeon General. Billings is seated second from right.*

One of Billings's many jobs in the surgeon general's office was to oversee the dismantling of military hospitals. He also toured all the Marine hospitals in the United States. He became an expert in hospital design and construction and, in 1870, wrote "A Report on Barracks and Hospitals."[15] Among other findings, he concluded that the spread of infections was an enormous problem in hospitals. "I believe," Billings wrote, "that no hospital should be constructed with a view to its being used as such for more than fifteen years."[16] The report, together with a second, "A Report on the Hygiene of the United States Army," published five years later, helped Billings establish himself as an authority on hygiene and sanitation, especially as it pertained to the military. Even more important for our purposes, the reports solidified Billings's credentials as a candidate to design Johns Hopkins Hospital a decade later. Around this time, Billings was also given responsibility for a modest collection of medical books known then as the Surgeon General's Library. For now, however, we want to turn our attention to his role in organizing the design and construction of the new Johns Hopkins Hospital.

Planning a State-of-the-Art Hospital

From the outset, Johns Hopkins envisioned endowing a hospital that embodied the most advanced principles in design and patient care. In a letter to the trustees on March 10, 1873, Hopkins wrote that the new hospital should "in construction and arrangement compare favorably with any other institution of like character in this country or in Europe."[17] The founder's instructions to the trustees further stipulated that the hospital shall admit "the indigent sick . . . without regard to sex, age, or color."[18] (We will see in a later chapter that the implementation of this equitable ideal soon became distorted by policies and prejudices, both internal and external to Hopkins.) In 1875, the trustees of Hopkins Hospital put out a call to five experts in hospital construction to submit proposals for the new institution. The proposals were to include plans for matters including construction, heating, ventilation, and administration of the new hospital along with diagrammatic sketches.[19]

Very much on the trustees' minds when they requested the experts' proposals was the new but urgent topic of so-called hospitalism, the rise in mortality observed in large hospitals due to confining many patients in crowded, unhygienic spaces.[20] Billings had previously dealt with the issue in his "A Report on Barracks and Hospitals" for the Surgeon General's Office. In fact, all of the proposals addressed hospitalism, either explicitly or implicitly, and several included similar design concepts.

Billings's plan was the one the trustees ultimately accepted. In his excellent article for the Maryland Historical Society, A. McGehee Harvey, a former physician-in-chief at Johns Hopkins, noted, "it was Billings the man, rather than his proposal, that influenced their final choice."[21] In Harvey's view, the trustees quickly recognized that they would need the expertise of a physician to resolve the many medical and scientific issues that were beyond their scope as laymen. They were especially attracted by Billings's deep knowledge and strong reputation in the fields of hygiene and hospital organization. His views on medical education also appealed to a group charged with constructing a new university-based model of physician training. Working as he did in the Library of the Surgeon General's Office in Washington, DC, Billings was, in addition, nearby in case he needed to be consulted. In short, he was "an ideal choice to assist the trustees in carrying out the will of Johns Hopkins."[22]

Billings's plan for the new Johns Hopkins Hospital created a pavilion layout with separate buildings to reduce the transmission of diseases. As he had in "A Report on Barracks and Hospitals," Billings contemplated a temporary plan for Hopkins in which the pavilions would be torn down every twelve to fifteen years to prevent the accumulation of pathogens, but this concept was never implemented. To better understand the innovative features of the plan Billings developed for Johns Hopkins Hospital, it is helpful to look at his design (figure 2.3). (The actual drawings that Billings submitted—and subsequently revised after his tour of European hospitals in 1876—were likely produced by an architect.)

Several forward-thinking features of Billings's design stand out, some of which are illustrated in the accompanying photographs from the Chesney Medical Archives (figures 2.4–2.6). The Common Wards were intentionally planned without elevators to prevent the flow of airborne pathogens from one story to another. Patients had to be carried up stairways to their beds.[23] (Billings's writings make it clear that he knew about the germ theory of disease but sometimes referred to miasma when describing the spread of infection.) To further reduce the transmission of infection, these stairways opened to the outside, reducing the direct exchange of air from one building to another.[24] On the inside, the wards were bright and designed to admit the maximum amount of sunlight. The ventilation grates visible under patient beds were also noteworthy (figure 2.4). Billings wanted to create as much airflow as possible to reduce the spread of pathogens.

Like the other patient buildings, the distinctive Octagon Ward was designed to expose patients to natural light. The interior view in the photograph reveals that every bed touched a window (figure 2.5). Though Billings had the foresight to prewire the hospital for electricity years before power was connected to the site, plumbing remained a challenge. Looking closely at the center of the ward in the same photo, it's clear that the sink drained into a bucket. Not everything in the new Johns Hopkins Hospital was completely modern!

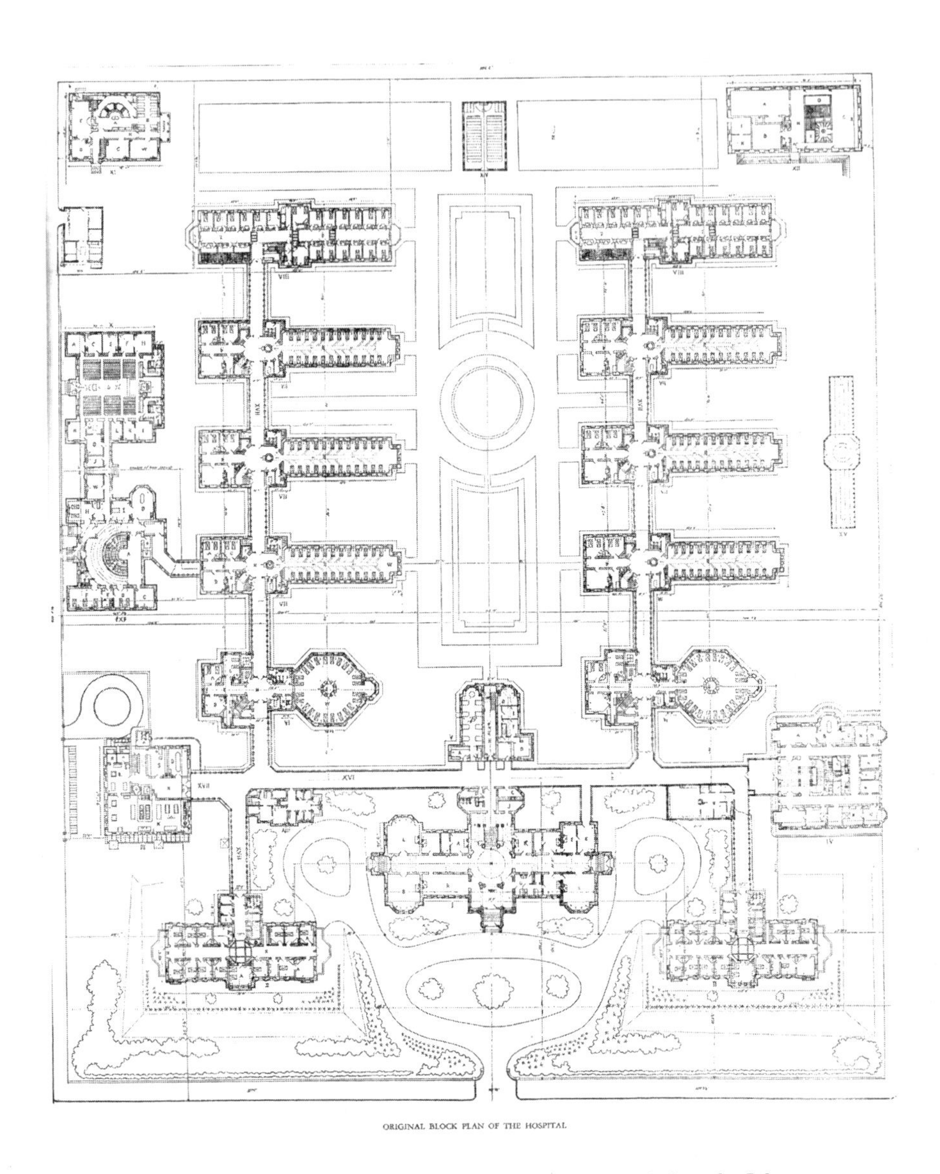

ORIGINAL BLOCK PLAN OF THE HOSPITAL

FIGURE 2.3. *Drawings, circa 1877, from the original plans for Johns Hopkins Hospital. Billings created a pavilion plan to reduce the transmission of infections. The main entrance, what is now called the Billings Administration Building, is just below center at six o'clock.*

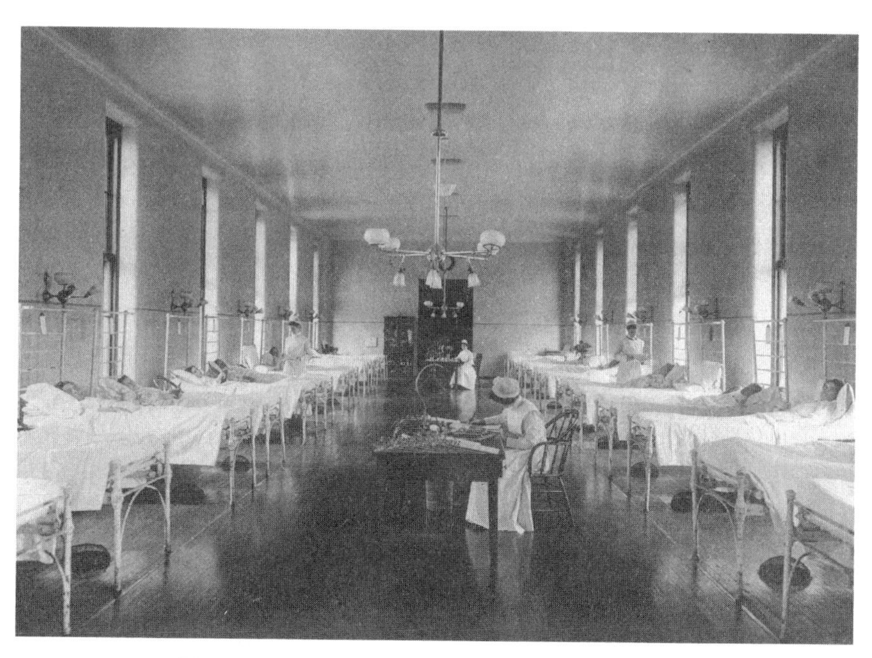

FIGURE 2.4. *The Common Ward. Ventilation grates on the floor under each bed were used to increase airflow, and many windows provided a light-filled interior.*

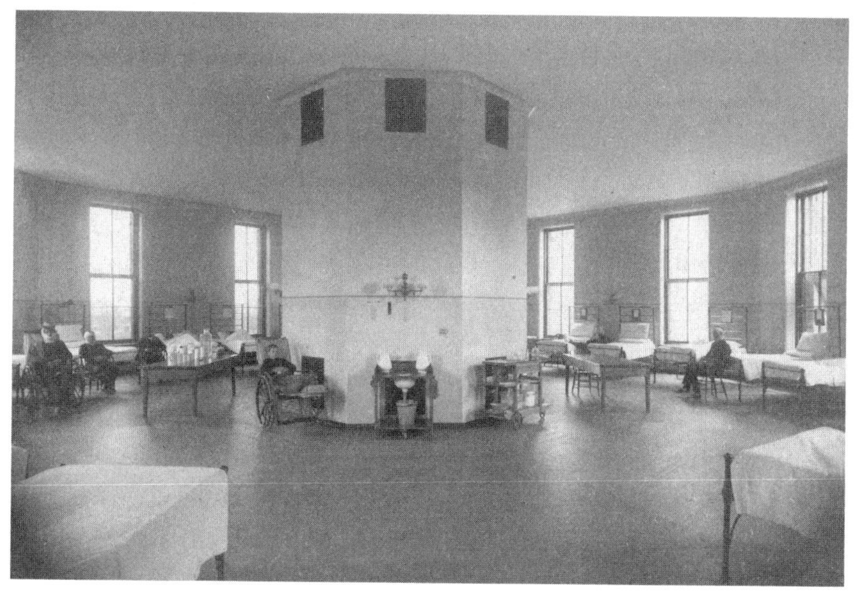

FIGURE 2.5. *Interior of the Octagon Ward. Every bed was adjacent to a window, though the absence of indoor plumbing meant that sinks in the ward drained into buckets, visible in the center of the photo.*

The Isolation Ward contained about twenty private rooms (figure 2.6). As the chimney stacks reveal, each room had its own fireplace and separate ventilation to promote airflow and reduce the transmission of pathogens. Remarkably, when the nurses in this unit were on duty, they were required to stay on the ward for several days at a time because of the fear that going and coming might spread infectious diseases.[25]

FIGURE 2.6. *Exterior view of the Isolation Ward in 1889. It housed about twenty private rooms. The chimney stacks rising above the roofline reveal that each room had its own ventilation to control the spread of pathogens.*

In addition to keeping uppermost in his mind Johns Hopkins's ambition to create a first-rate hospital, Billings also paid close attention to the founder's insistence that that "the institution shall ultimately form a part of the medical school of that university for which I have made ample provision by my will."[26] Billings brought these guiding principles to life, transforming them into the fundamental mission of Johns Hopkins Medicine: the integration of rigorous medical education, compassionate patient care, and pioneering scientific research.[27]

Billings understood the power of evidence and ideas discovered by a new breed of physician-scientists. He singled out the pathology laboratory during his remarks at the hospital's opening ceremonies in 1889, emphasizing, "Upon the results obtained in that laboratory may yet depend the saving of many lives, the relief of unspeakable agony, [and] the warding off of pestilence from the city."[28]

In contrast to other leaders in American medicine of his era, Billings envisioned that research would be a central component of medical training at Hopkins. In his essay "Hospital Construction and Organization," Billings wrote that the medical school should aspire to produce not only a well-trained physician capable of making accurate diagnoses, but also one who is able "to investigate for himself."[29] He argued that Hopkins must seek "to give the world *men* [italics ours] who can not only sail by the old charts, but who can make new and better ones for the use of others"[30] Billings's progressive thinking about medical research and education remains true today, but his "boys club" language, referring to physicians as men by definition, reveals that he was a captive of his era's chauvinistic culture.

Billings was eager to bring medical education inside the walls of Hopkins Hospital and to make the wards an extension of the classroom. Along with caring for the poor and advancing scientific research, Billings emphasized in his speech that the hospital "should provide the means of giving medical instruction, for the sake of the sick in the institution as well as those out of it."[31] In Billings's view, "The sooner he [the student] can begin to profitably receive instruction by the bedside of the sick . . . the better." He was quick to add, "Nothing can take the place of this; if it be not obtained before graduation, when errors can be prevented by the teacher, it must be obtained afterwards at the expense of the first patients who present themselves."[32] While the value of teaching medical students in a hospital may seem obvious today, the idea, although widely accepted in Europe, was a novel one at the time in the United States.

Finding the School of Medicine's Founding Four

Billings also was instrumental in recruiting the leaders of the new medical school. Initially, he hired William Welch, the school's first dean and first head of the Pathology Department. In 1876, after Billings had accepted the commission to design Hopkins Hospital but more than a dozen years before the institution opened, Billings traveled to Europe to study the design and organization of hospitals, and, in so doing, he also brought back their science-based philosophy and approach to medicine. On November 7, Billings visited the laboratory of the German scientist Ernst Wagner in Leipzig, Germany. While touring the lab he met Welch, who was studying there. That evening, Billings and Welch met for beer in Auerbach's Keller—the very place where Faust was reputed to have made his bargain with the devil.[33] Despite the inauspicious setting, the two physicians shared a common vision of the future of medicine, and their meeting set in motion the plan to bring Welch to Johns Hopkins.

Billings also recruited William Osler, the first chief of medicine, more than a decade later. In Osler's telling, "Early in the spring of 1889 he came to my rooms, Walnut Street, Philadelphia . . . Without sitting down, he asked me abruptly: 'Will you take charge of the Medical Department of the Johns Hopkins Hospital?' Without a moment's hesitation I answered: 'Yes.' 'See Welch about the details; we are to open very soon. I am very busy today, good morning,' and he was off, having been in my room not more than a couple of minutes."[34] Not surprisingly for a man as driven as Billings, he didn't hesitate, when the time came, to put his new hires to work recruiting other first-rate physicians. Welch was instrumental in recruiting William Stewart Halsted, the first chair of surgery at Hopkins, and Osler helped recruit Howard Atwood Kelly, the first head of gynecology and obstetrics.

In the decade and a half between his appointment by the trustees at Hopkins in 1876 and the opening of the hospital in 1889, Billings had fundamentally transformed the concept of an American medical institution. He had designed a new, state-of-the-art hospital incorporating the latest ideas about patient care and infection control. He had established research and patient-centered instruction of medical students as the twin foundations of Hopkins Hospital's mission in caring for the sick. And he had recruited the best and the brightest candidates for the founding faculty of the School of Medicine and set them on their groundbreaking path in medical discovery and physician training. Billings's own words best describe his ultimate ambition: "Let us hope that before the last sands have run out from beneath the feet of the years of the nineteenth century it [Hopkins] will have become a model of its kind."[35] The centennial history of Johns Hopkins Medicine, indeed titled *A Model of Its Kind*, was published in 1989 and took its inspiration from Billings. The two-volume account confirmed Billings's historic achievements and the rightness of his vision.[36]

A Universal Medical Bibliography

Even with his signal accomplishments designing Hopkins Hospital, developing its philosophy, and recruiting many of its early leaders, Billings wasn't done. Amazingly, in a role concurrent with his work at Hopkins, he guided the building of the Library of the Office of the Surgeon General. As noted earlier, Billings was put in charge of a modest medical literature collection when he joined the surgeon general's office. Almost from the start, and perhaps influenced by his arduous experience as a medical student gathering sources for his article about epilepsy, Billings viewed his task as not only adding books and articles but also increasing the ability of physicians and other medical researchers to find and use these materials. By 1883, Billings had declared, "We are endeavoring to make this Library complete in Medical Literature—in order that there may be one collection in the world to which a person seeking information can apply with a reasonable certainty that he can find in it all that has ever been published relative to any medical subject."[37]

In 1867, before Billings took on the project, the library had 1,800 volumes. Billings personally selected and purchased many of the books and journals, and by 1895, under his leadership, it had grown fiftyfold to 90,000 volumes.[38] The collection by then included the latest periodicals and allowed civilians as well as military personnel to access its resources. What had been the Army Medical Library of the Surgeon General's Office later became the Armed Forces Medical Library and eventually the National Library of Medicine, one more transformative institution in American medicine built by John Shaw Billings.

As the collection rapidly expanded, the need for a system of organization grew as well. *Information storage and retrieval* is a distinctly modern term, but it aptly describes the ahead-of-its-time approach that Billings took in collecting, organizing, and indexing the Surgeon General's Library. As Charles Coffin Jewett, then assistant secretary in charge of the library at the Smithsonian Institution, put it, "A library without a catalogue is like a body without an eye."[39] Billings's solution was to review, mostly by himself, tens of thousands of journals and books and create an alphabetical index-catalogue of all authors and subjects in the medical literature. Billings produced his first monumental volume ("A to Berlinski") of the *Index-Catalogue of the Library of the Surgeon-General's Office, United States Army* in 1880 (figure 2.7). He published a new volume each year through 1895. The volumes averaged more than one thousand pages each. In total, the *Index-Catalogue* referenced more than 170,000 books and pamphlets, and over 500,000 journal articles.[40] The updates to the *Index-Catalogue* became the *Index Medicus* and, ultimately, today's powerful PubMed database, which, according to the National Library of Medicine, the direct descendant of Billings's surgeon general's library, comprises "more than 32 million citations for biomedical literature."[41]

Billings created an index card for each article and punched a hole in a particular location to designate the subject matter covered. By subsequently putting a wire about the diameter of the end of a paperclip through a stack of cards, every card with a hole in the same position could be pulled, and all of the information on a given subject—say, "pancreas"—could easily be drawn together.

Osler called the *Index-Catalogue* "one of the most stupendous bibliographical works ever undertaken in any field . . . [It furnishes] to the world a universal medical bibliography from the earliest times. It will ever remain a monument . . . to the enterprise, energy and care of Dr. Billings."[42] Similarly, Robert Koch, the discoverer of the pathogen that causes tuberculosis, wrote, "You have created for future medical writers a quite indispensable tool."[43] Billings was widely acclaimed for the *Index-Catalogue.* After his speech before the International Medical Congress titled "Our Medical Literature," the audience "shouted and cheered until they were hoarse."[44] The clamorous reception sometimes propelled him into interesting social situations. On August 3, 1881, Billings was invited to a luncheon in his honor in London hosted by the renowned British surgeon Sir James Paget. He was seated with, among other luminaries, Charles Darwin, the prince of Wales, and, immediately to Billings's right—in a weird twist—Dr. William Gull, whom some people think was the notorious Jack the Ripper. Quite a group of dining companions!

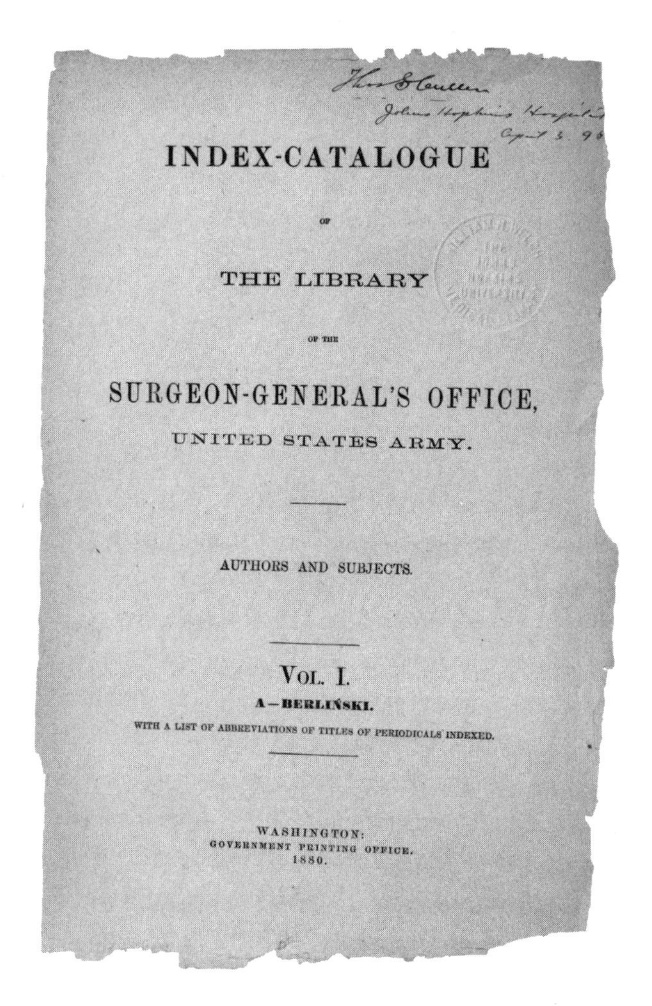

INDEX-CATALOGUE

OF

THE LIBRARY

OF THE

SURGEON-GENERAL'S OFFICE,

UNITED STATES ARMY.

AUTHORS AND SUBJECTS.

VOL. I.

A—BERLIŃSKI.

WITH A LIST OF ABBREVIATIONS OF TITLES OF PERIODICALS INDEXED.

WASHINGTON:
GOVERNMENT PRINTING OFFICE.
1880.

FIGURE 2.7. *The title page from volume 1 of Billings's monumental index-catalogue of medical information. The volumes averaged more than one thousand pages. Billings published a new volume each year between 1880 and 1895. The updates eventually evolved into today's PubMed database. (Photograph by Norman Barker.)*

With his strong interest in information and information management, Billings naturally gravitated to the application of statistics to public health. In his contribution to a book about hygiene and public health published in 1873, Billings wrote, "Sanitary measures, to be effective, should be carried out at those times when most people see no special cause for anxiety, and often, therefore, appear to involve unnecessary worry and expense."[45] (If we extend Billings's focus in his 1873 article to public health measures generally, his words contain rich advice relevant to the COVID-19 pandemic, for which the health care system was woefully unprepared.) An early proponent of a proactive approach to public health, Billings was appointed as the first director of the Institute of Hygiene at the University of Pennsylvania.[46]

The 1880 Census and an Ingenious Punch Card System

Peers in government who faced the challenge of collecting and organizing large amounts of information took notice of Billings's organizational brilliance. Together with the businessman and statistician Herman Hollerith, Billings was invited to participate in planning the 1880 census. He introduced Hollerith to the punch card system for sorting and classifying different categories of information he had developed for the Library of the Surgeon General's Office (figure 2.8). Billings wrote that one could "record many of the data on . . . cards by punching slots or holes in them in such a way that the several enumerations required could be made by electrical counting or by distributing the cards by machinery."[47]

In his *History of the Electrical Tabulating Machine*, Hollerith recalled, "While engaged in work in the Tenth Census, that of 1880, my attention was called by Dr. Billings to the need of some mechanical device for facilitating the compilation of population and similar statistics."[48] Eventually Hollerith refined this strategy for sorting and cataloguing information, turning it into the Tabulating Machine Company, a business that manufactured electrical calculating machines. Hollerith sold his company in 1911 for 1.2 million dollars and, after many iterations, it became International Business Machines (IBM).

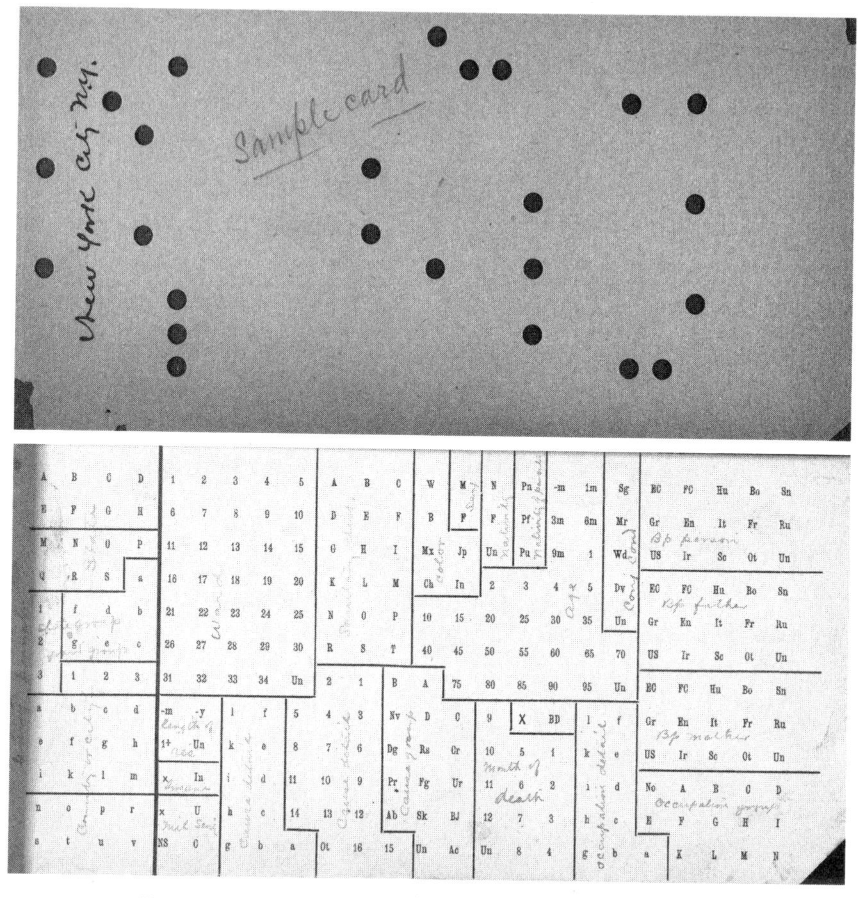

FIGURE 2.8. *Punch cards from the Billings collection that were likely used in the US census. The holes punched into a card (top) designated features, such as age, occupation, and country of origin, as shown on the "decoder card" (bottom). Billings shared his punch card idea with Herman Hollerith, who started an electrical tabulating company that eventually became IBM.*

Many a talented individual would have been inclined to take it easy if they came even close to rivaling Billings's highly accomplished military, medical, or governmental career—let alone all three—but he was still eager to take on new challenges. In 1896, Billings assumed the directorship of the New York Public Library. In that role, he guided the merger of the Astor Library, the Lenox Library, and the Tilden Trust into what we now know as the New York Public Library system. He helped create author-subject catalogues and he increased the availability of standard references by arranging them around the perimeter of the main reading room.[49] Billings convinced Andrew Carnegie to donate $5.2 million to build the beautiful main branch of the New York Public Library that stands at the corner of Fifth Avenue and 42nd Street with its immense sculpted lions guarding the entrance. Billings was responsible for that building and personally helped design the majestic Rose Main Reading Room (figure 2.9). At the New York Public Library, as at so many other points in his career, Billings's pivotal contributions lay in his ability to marshal a unique combination of technical expertise, organizational leadership, and, where necessary, personal suasion. Interestingly, those who knew him well noted that Billings was not a particularly affable or gregarious character.

FIGURE 2.9. *The Rose Main Reading Room of the New York Public Library. In addition to guiding the creation of the New York Public Library system, Billings took the lead in designing this elegant space. (Photo by David Iliff, license: CC BY-SA 3.0.)*

Billings's Legacy

John Shaw Billings died on March 11, 1913, and was buried at Arlington National Cemetery. William Stewart Halsted, the first chair of surgery at Hopkins, was an honorary pallbearer at the funeral. The list of Billings's life achievements is long and quite amazing. He was a Civil War hero. He designed Johns Hopkins Hospital. He established the early philosophy of the School of Medicine and recruited several of its founding faculty members. He built the National Medical Library. He developed the *Index-Catalogue*. He led the department of public health at a major university. He was instrumental in the development of modern computer technology. And he created the New York Public Library system.

When we think of John Shaw Billings, we think, perhaps above all, of science tirelessly applied to medicine and the wider public good. His biographer Carleton Chapman highlights Billings's "indomitable pursuit of his goal."[50] Billings himself wrote this in the *Boston Medical and Surgical Journal* in 1886: "The best life . . . is that which is devoted to the helping of others, which is unselfish, not stained by envy or jealousy, and which has as its main pleasure and spring of action the desire of making other lives more pleasant, of bringing light into dark places, of helping humanity."[51]

Of course, there can be a heavy personal and familial price to pay for this kind of selflessness. One of his biographers reported that Billings chose not to attend his mother's funeral in 1898 because it conflicted with a business engagement.[52] John Shaw Billings's son is said to have remarked, "I would not . . . lead my father's life for twice his name and fame . . . I have always hated my father's life . . . hated it fiercely and vindictively . . . for what it deprived him of."[53]

We will perhaps never resolve the moral dilemma of how to comprehend the lives of people like Billings who gave themselves so fully to humanity that they cheated the very humans who loved them most. But one thing is certain: John Shaw Billings was an extraordinary builder. He was a builder not of products or places, though his name is on the iconic domed administration building at Johns Hopkins Hospital, but a builder of entire systems that solved fundamental challenges with ingenious solutions far ahead of the times. Billings invented the *Index-Catalogue* that finally made it possible for medical researchers to stand on the shoulders of scientific investigators from across the globe who had come before them. The punch card system he conceived and passed on to Herman Hollerith freed the US Census from the arduous and largely meaningless hand counting of demographic data and brought it into the age of powerful statistical analysis.

Certainly for us, with our focus on American medicine, Billings's crowning achievement was designing and bringing into being the Johns Hopkins medical institutions. He laid out the plans and oversaw the initial construction of a hospital based on the latest hygienic and sanitation principles. Billings also articulated a vision of patient care that harnessed the twin engines of research and bedside teaching to improve the healing process. Most important of all, Billings recruited physician-scientists of the first order to the new Johns Hopkins and made the discovery of new medical knowledge and treatment methods the beating heart of the new hospital. His dream as the architect of a brand-new type of medical institution, as in the case of virtually all of his major endeavors, was at once visionary and rooted in science. As Billings laid it out in one of his first reports to the trustees on the organization of the new Johns Hopkins: "This Hospital, should advance our knowledge of the causes, symptoms and pathology of diseases, and methods of treatment, so that its good work shall not be confined to the city of Baltimore, or the State of Maryland, but shall in part consist in furnishing more knowledge of disease and more power to control it, for the benifit of the sick and afflicted of all countries and of all future time."[54]

We, who live in that future time, can testify to the prophetic nature of Billings's dream.

ɔLOR PLATE I, ABOVE: Billings Administration Building, circa 1889, the year the Johns Hopkins Hospital ɔened.

COLOR PLATE 2, ABOVE: Perhaps the most famous painting in American medicine, *The Four Doctors* by John Singer Sargent depicts four of the founding physicians of the Johns Hopkins University School of Medicine: William Welch, William Halsted, William Osler, and Howard Kelly (left to right). (*The Four Doctors* by John Singer Sargent, 1906, oil on canvas.)

COLOR PLATE 3, RIGHT: John Singer Sargent's portrait of Mary Elizabeth Garrett. (*Mary Elizabeth Garrett* by John Singer Sargent, 1904, oil on canvas.)

COLOR PLATE 4, RIGHT: Portrait of John Shaw Billings. (*John Shaw Billings* by Bradley Stevens after Cecilia Beaux, 1989, oil on canvas.)

COLOR PLATE 5, LEFT: Cover of the April 14, 1930, issue of *Time* magazine. At the age of eighty, William Welch was hailed as the "Dean of American Medicine" who had built Hopkins into a "model of its kind."

COLOR PLATE 6, LEFT: Portrait of Welch. (*William Henry Welch* by Thomas Corner, 1920, oil on canvas.)

COLOR PLATE 7, RIGHT: This portrait of Sir William Osler hangs in the office of the director of the Department of Medicine at Johns Hopkins Hospital. (*Sir William Osler* by Thomas Corner, 1905, oil on canvas.)

COLOR PLATE 8, ABOVE: *The Gross Clinic*, Thomas Eakins's 1875 portrait of Dr. Samuel D. Gross. The surgeon is operating bare-handed, blood visible on his fingers, and the patient's mother cowers in the background. (*The Gross Clinic* by Thomas Eakins, oil on canvas.)

COLOR PLATE 9, LEFT: Portrait of William Stewart Halsted painted posthumously. (*William Stewart Halsted* by Thomas Corner, 1936, oil on canvas.)

COLOR PLATE 10, ABOVE: A portion of the stained glass *Sacrifice*, in the National Cathedral in Washington, DC. Note the mosquito in the upper right and Lazear applying a test tube containing the disease-carrying insect to his arm in the center.

COLOR PLATE 11, ABOVE: Brödel created beautiful illustrations, including this one of Lake Ahmic in Ontario, on the white undersurfaces of bracket fungi using a fine-pointed tool. (Photograph by Ralph Hruban).

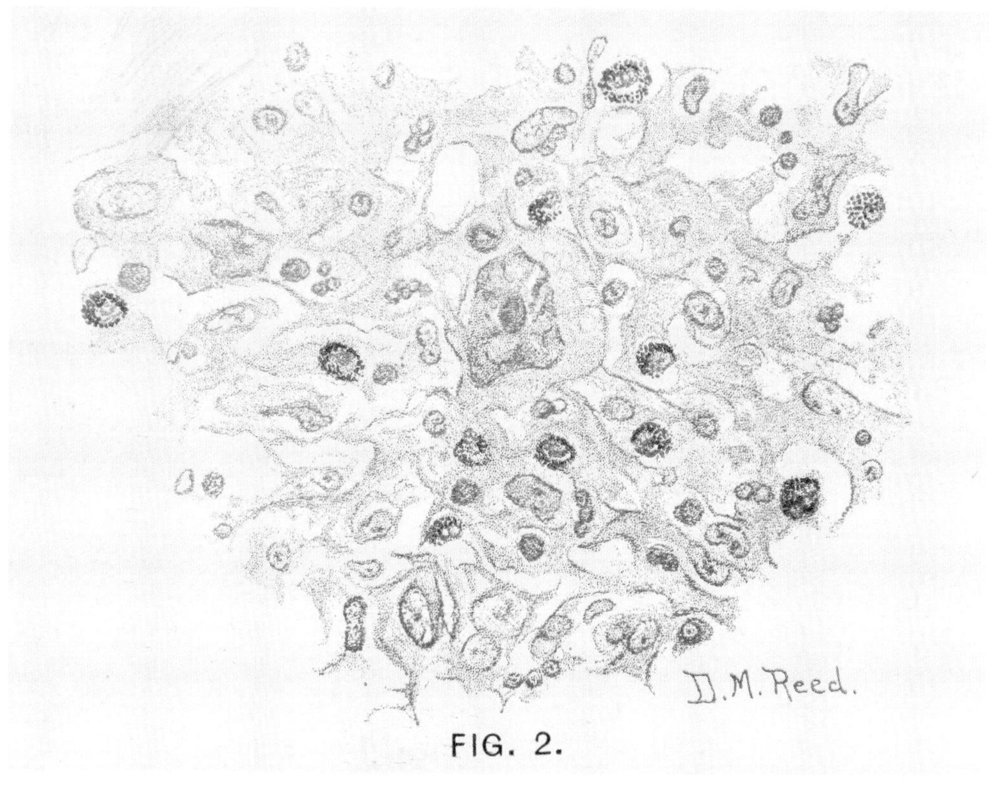

COLOR PLATE 12, ABOVE: Reed illustrated her manuscript on Hodgkin disease in her own hand. (From D. Reed, "On the Pathological Changes in Hodgkin's Disease, with Special Reference to Its Relation to Tuberculosis," *Johns Hopkins Hospital Reports* 10 [1902]: 133–96.)

COLOR PLATE 13, LEFT: Jamie Wyeth's portrait of Helen Taussig. Wyeth would later say, "She would stare right at me intently with those blue eyes. It was so amazing." (*Helen Brooke Taussig* by Jamie Wyeth, 1963, oil on canvas.)

COLOR PLATE 14, RIGHT: Portrait of Vivien Thomas. (*Vivien Theodore Thomas* by Bob Gee, 1969, oil on canvas.)

3

WILLIAM HENRY WELCH

Medicine through the Lens of Science

With her remarkable vision and shrewd philanthropy, Mary Elizabeth Garrett stood up to Hopkins president Daniel Gilman, who had denied her a college education, and set the School of Medicine on its historic path toward rigorous standards and fair treatment of women. John Shaw Billings, a veteran of the bloody battles at Chancellorsville and Gettysburg and an expert in the construction and organization of military hospitals, designed Johns Hopkins Hospital, recruited several of its founding doctors, and shaped an identity grounded in teaching and research. It remained for William Henry Welch, Billings's initial recruit and the first dean of the medical school, to begin the work of implementing the new scientific approach to physician training and patient care that would revolutionize American medicine.

There can be little doubt that Welch possessed exceptional qualifications for the task. The son of a genial New England country doctor who had a caring bedside manner, classically educated in literature and the arts at Yale, and having spent time abroad with some of Europe's most renowned physician-researchers, Welch was in a unique position to advance science-based medicine in America. He certainly had the intellectual firepower. His Hopkins colleague William Osler, a legend in his own right whom we will meet in the next chapter, said of Welch, "In addition to a three-story intellect [he] has an attic on top."[1] Besides a probing mind, Welch had another kind of intelligence—something we might describe today as emotional IQ—that gave him an extraordinary ability to influence his colleagues and peers.

A Sad State of Affairs: Physician Training as the Twentieth Century Opened

Despite the new model of medical education and practice established when the Johns Hopkins School of Medicine opened in 1893, the transformation of American medicine was slow and piecemeal. It wasn't until 1901 that Harvard followed Hopkins's lead in requiring medical students to have a college degree.[2] As late as 1907, the American Medical Association's newly formed Council on Medical Education published a damning yet confidential (for fear of antagonizing its members) report on 162 medical schools in the United States and Canada. The report found the majority of medical schools were not affiliated with any university or hospital, lacked any admission standards, and were owned by faculty who took their salaries from student tuition. Worse yet, as John Barry points out in *The Great Influenza*, "One school had graduated 105 'doctors' in 1905, none of whom had completed any laboratory work whatsoever; they had not dissected a single cadaver, nor had they seen a single patient."[3]

A second report on American medical education issued publicly in 1910, this time by the respected Carnegie Foundation for the Advancement of Teaching, was the game-changer.[4] The Carnegie Foundation evaluators noted that two schools in Arkansas were "without a single redeeming feature," and three medical schools in Texas were "without resources, without ideals, without facilities."[5] The report recommended that of the roughly 155 medical schools in the United States and Canada, 80 percent be closed.[6] This was the inauspicious environment Welch faced in the early years of his career at Hopkins, hardly conducive to embodying and spreading Garrett and Billings's dreams of creating a new generation of rigorously trained, scientifically grounded physicians. Despite the obstacles, Welch proved himself uniquely suited to the task.

The Son of a Country Doctor

William Henry Welch was born on April 8, 1850, in Norfolk, Connecticut. His mother died when he was only six months old, and he was raised by his grandmother. His father was a country doctor using old-time remedies: For a rattlesnake bite, the senior Dr. Welch prescribed "copper filings from a large copper penny of the reign of Queen Ann[e], dissolved in vinegar, and immediately applied to the wound."[7] Fortunately, New England rattlesnakes were small and rarely lethal.

In 1866, at the age of sixteen, William Welch entered Yale to study the classics. He focused his work on religion, posing a question in his graduating essay that offered no hint of his later interest in science: Did it not "make life hopeless and not worth living" to think of the world "as a vast machine unguided by a God of justice?"[8] Welch was a brilliant student, graduating third in his class in 1870. But as a classmate noted, despite narrowly missing being the valedictorian, "there was about him no suggestion of the 'dig,' that bête noir of the average underclassman. His acquisition of knowledge was effortless; he seemed to absorb it. It was as easy as breathing."[9] That easy affability and fluid intelligence would set Welch apart throughout his life.

Welch's ambition after graduation was to teach Greek at Yale, but there were no positions available. For the next academic year, he taught Latin and German at a newly organized academy in Norwich, New York, before turning in quiet desperation to his father's profession. "Welch's future looked so black and empty," wrote his biographer and former student Simon Flexner, that "he listened to his father's voice as it urged that he follow the family tradition and go into medicine."[10] Late in 1871, Welch apprenticed with his father.

The following year, he entered the College of Physicians and Surgeons, now Columbia University's medical school, in New York City. The school, like most other medical schools, was strictly a business enterprise, with no admission requirements other than being able to pay tuition. Welch had only one examination to pass at the end of his three years of training there. He described it as "the easiest . . . I ever entered since leaving boarding school."[11]

Graduating in 1875, Welch went on to intern in pathology at New York's Bellevue Hospital, placing him at the pinnacle of that discipline, which was still in its infancy in America. Within a year, he did what many of his more ambitious and scientifically minded physician peers did and traveled to Europe. As Barry puts it in his book on the first global pandemic, "Little medical research was being done in America . . . but . . . in Europe science was marching from advance to advance, breakthrough to breakthrough."[12] Among colleagues at home, Welch had a reputation for his extraordinary knowledge. In Europe he was refused acceptance into Friedrich Daniel von Recklinghausen's laboratory because he knew too little.[13] Nonetheless, he managed to study at universities in Strasbourg, Leipzig, Breslau, Vienna, and Berlin. He eagerly absorbed the European model of science applied to medicine, which he energetically promoted when he returned to America.

We noted earlier that while Welch was in Europe, he met John Shaw Billings at Auerbach's Keller in Leipzig. Billings was in Europe to study the design of hospitals in order to complete his plan for Johns Hopkins. Sitting beneath a painting of Faust making his pact with the devil, the two physicians discussed quite a different scheme—namely their shared ambition to bring scientific medicine to America. Though Billings did not offer Welch a position at Hopkins at the time, he apparently was sufficiently impressed that he told Hopkins trustee Francis King, who had traveled with him to Leipzig, "that young man should be, in my opinion, one of the first men to be secured when the time came to begin the medical school."[14] Welch returned to the United States in 1878. The College of Physicians and Surgeons had no job to offer him, so he went back to Bellevue, where he spent the next six years teaching pathological anatomy and general pathology at the hospital's medical school.

At Bellevue, Welch was provided with three rooms for his research fitted out with kitchen tables. The hospital had spent, Welch wrote mournfully, "fully twenty-five dollars in equipping the laboratory."[15] Even with the obstacles in his path, Welch managed to design America's first laboratory course in pathology and began his research. Medical students from all over New York City and surrounding areas flocked to hear him lecture. Despite this growing recognition, Welch despaired of the sorry state of medicine in the United States. In a letter to his sister, he wrote, "I sometimes feel rather blue when I look ahead and see that I am not going to be able to realize my aspirations in life . . . There is no opportunity in this country, and it seems improbable that there ever will be . . . there is no opportunity for, no appreciation of, no demand for that kind of [research] here."[16]

Though Welch's prospects in those years at Bellevue might have seemed bleak, there were lighter moments that revealed his lifelong sense of fun. He often caught frogs for teaching and once carried a box full of them aboard a New York train. Soon the car was filled with the sounds of croaking amphibians. Welch, with the box neatly tucked under his seat, pretended to be as surprised as his fellow passengers.

The Summons from Hopkins

In the spring of 1884, five years before the opening of Johns Hopkins Hospital and nine years before the School of Medicine enrolled its first class, university president Daniel Gilman offered Welch a faculty position as chief of pathology. At that time—as it is today—pathology was on the leading edge of scientific medicine. The top job in the yet-to-be-opened but much heralded medical school was a plum appointment. Welch avowed, “Such great things are expected of the medical faculty at the Johns Hopkins in the way of achievement and of reform.”[17]

The move to Hopkins was not without its complications—complications that, as we shall see, shed light on Welch’s inner life. While in New York, Welch had formed a deep and perhaps intimate relationship with Frederic Dennis, Welch’s former boarding school roommate. The son of a railroad baron and a physician himself, who like Welch had studied in Europe, Dennis had sometimes paved the way for Welch in New York with his family connections and occasionally subsidized Welch financially.

When the offer from Hopkins came, Dennis insisted that Welch turn it down, telling any acquaintances who would listen, “If he goes, he must sacrifice the friendships of a life and isolate himself in a provincial city among untried and inexperienced persons.”[18] Welch wavered for a time, and Dennis and their mutual friends orchestrated a behind-the-scenes campaign to get Bellevue to build Welch a lab that would match any at Hopkins. In the end, Welch accepted the offer from Hopkins, concluding it was “undoubtedly the best opportunity in this country.”[19] Still troubled but resolute, Welch wrote his stepmother, declaring he grieved that “a life-long friendship should thus come to an end.”[20]

When Welch agreed to take up his new position at Hopkins on March 31, 1884, he was thirty-three years old. He became the eighth full professor appointed at the university and the first in the School of Medicine. “I can develop my field in Baltimore unhampered by traditions,” he announced.[21] Prior to this, not a single medical school in the United States had a laboratory dedicated to the basic science of pathology, and no American physician had ever made a living by teaching and scientific work alone.[22]

Welch as a Teacher and as Head of the Pathology Lab

Welch was known as a great teacher, possibly because he didn’t give priority to textbooks (figure 3.1). Instead, he focused on original sources and having students do their own research. Laboratories took the place of the uninspiring lectures traditional in American medical schools, and experiments replaced rote precepts. Milton Winternitz, an early Hopkins medical student who went on to become the dean of Yale Medical School, captured Welch's distracted but uniquely engaging style as a teacher: "[He] was quizzing the class and he rubbed his vest pocket against the table and a handful of sulphur matches ignited. He didn't stop talking but just took the matches out of his pocket and put them on the table, while we watched in amazement."[23]

FIGURE 3.1. *Welch at the chalkboard in 1903. "Popsy," as he was called, was an inspiring teacher.*

Gilman's offer to Welch included provisions for further study and travel in addition to funding to set up the new Pathology Laboratory. By September 1884, Welch had returned to Europe, where he investigated the organization and instrumentation of more than a half dozen leading medical laboratories, including those of Robert Koch, the discoverer of the pathogen that causes tuberculosis and the famed developer of the germ theory of disease that was transforming medicine, as well as the laboratory of the world-renowned Louis Pasteur. He also used the yearlong trip to collect bacterial cultures and purchase equipment for his new laboratory at Hopkins.

Welch departed for Baltimore in August 1885. When he arrived, he set about creating a new type of pathology laboratory, one that was permeated with what one observer called "a general spirit of research" and "a fine esprit de corps. Clinicians and laboratory men were living together under circumstances of delightful intimacy"[24] (figure 3.2). Women were welcome in the Pathology Laboratory. As Lilian Welsh, a pioneering female physician, described it in her memoir, "Anyone, man or woman . . . was made welcome to share in the most stimulating medical companionship that possibly this country has ever known."[25] Democratic principles prevailed, with Simon Flexner, Welch's former student at Hopkins, remarking, "One unusual feature of Welch's laboratory was the entire lack of distinction between younger and the more experienced or highly trained men. Opportunity to grasp a theme, opportunities to pursue it, were open equally to both."[26] Remarkably, for a time when anti-Semitism was on the rise, especially in higher education, Welch mentored Simon Flexner and Milton Winternitz, who were Jewish, and helped to further opportunities for them throughout their respective careers.

Welch did much more than just set the tone for productive research in the Pathology Laboratory. He had the ability to hear someone describe a scientific problem and immediately define the critical points and key experiments that had to be done. By design, research was carried out at the same time as teaching. Students found themselves swept up in the charged intellectual atmosphere created by working on fresh science. In Welch's view, that science needed to be basic science that uncovered fundamental biological and physiological principles. He argued, "The most important discoveries in science, come not from those who make utility their guiding principle, but from the investigators of truth for its own sake."[27]

FIGURE 3.2. *Welch at the microscope in 1905. More than anyone, he brought science to the field of medicine in America.*

In contrast to students at other medical laboratories, Welch's students, even those who were quite junior, often undertook research on their own. As a result, many independent investigators were created. This method was so effective and Welch's prestige so great that his students were often offered the best positions. Quite a few of his trainees went on to become directors of departments and deans at other institutions.[28] One medical historian called Welch's pathology laboratory "a Mecca for young men from various parts of the country" and proceeded to reel off a list of the distinguished physicians who received their training from Welch.[29] The cartoon in figure 3.3 was penned by famed medical illustrator Max Brödel, whom we will meet in chapter 7. Wryly titled *Some Welch Rabbits*, it depicts more than a dozen prominent physician-researchers who had been Welch's students and had gone on to lead departments, hospitals, or universities across the country. The cartoon gives some indication of Welch's influence. His message about rigorous, evidence-based medicine rapidly spread throughout America.

FIGURE 3.3. Some Welch Rabbits *by Max Brödel. Many of Welch's trainees would go on to leadership positions in medical schools and hospitals across the country.*

A Natural Leader

Welch's keen intelligence and personal charisma made him an obvious choice for leadership roles (figure 3.4). At Hopkins, he was not only the first director of the Department of Pathology, but he went on to become the first dean of the School of Medicine and the first dean of the School of Hygiene and Public Health. In the latter role, Welch had an impact well beyond his academic duties. In fact, the *Baltimore Sun* credited Welch with playing a leading role in transforming the city "from a cesspool of outhouses and wells in the backyard of every dwelling to a modern sanitary city."[30] Late in his storied career at Hopkins, Welch was named the first director of the Institute for the History of Medicine.

FIGURE 3.4. *Welch at the age of forty-three, about the time the School of Medicine opened in 1893.*

Part of Welch's appeal as a leader was, in fact, his ability to recruit top talent. In 1893, he lured the renowned anatomist and embryologist Franklin Mall, who had been one of Welch's early postgraduate fellows but had left Hopkins and eventually became a professor of anatomy at the University of Chicago. Despite Chicago's offer of a considerable increase in salary, Mall accepted Hopkins's invitation to return, telling Welch, "[I] shall cast my lot with Hopkins . . . I consider you the greatest attraction. You make the opportunities."[31] Welch was equally successful—under more trying circumstances, as we shall see—in recruiting William Halsted, the great surgeon and another of our ten transformative figures, whom he first met at Bellevue in the late 1870s.

Not only at Hopkins but on the national stage as well, Welch emerged as the leading figure in American medicine. In rapid succession, he was named president of the Association of American Physicians, president of the American Association for the Advancement of Science, president of the American Medical Association, and president of the National Academy of Sciences.[32] Welch also founded the *Journal of Experimental Medicine*.[33]

Ironically, by 1900, with his multiple and seemingly endless roles and responsibilities, Welch had all but abandoned his own scientific investigations. If his eminence was not based on his research and publications, what did it rest on? Welch's colleague William Osler, Hopkins's first chief of medicine, put it best when he wrote in his secret diary: "By far the ablest man of the group was the professor of pathology, William H. Welch . . . A more unselfish man never lived—his time, his brains, his books, his purse, were at the command of all. He was the most even-balanced soul I have ever met—never perturbed, always in good humor, and always ready to look on the bright side of things. I never heard him pass a harsh judgment on any one. He had strong convictions and never was afraid of expressing his opinion. One always knew where he stood on any debatable question, on which he generally said the last word."[34] When disagreements surfaced about medical matters, Welch's leadership style as dean, according to one of his colleagues, was to avoid "ex cathedra" pronouncements and give ample credit to another person's position, while stating his own in a pleasant but persuasive way that avoided emotional confrontation.[35]

Welch's impact on medical training and practice didn't stop at America's shores. During his lifetime he was awarded many honorary citations, including the Order of the Royal Crown (Germany), the Order of the Rising Sun (Japan), and the Plaque for Public Health Service (United Kingdom). When the Rockefeller Foundation wanted to bring Western-style medicine to China, its top officials called on Welch to lead the 1915 commission to China that laid the groundwork for Peking Union Medical College (figure 3.5). The school opened in 1919 with the goal of making it a comprehensive institution having its own hospital, teaching, and laboratory facilities set up along modern lines. Welch himself expressed the hope that Peking Union Medical College, and others like it that were based on Hopkins's philosophy of research and training, would "serve as models for the country."[36] In 2017, coauthor Ralph Hruban had the honor and privilege of visiting Beijing to celebrate the school's centennial.

FIGURE 3.5. *Welch led the 1915 Rockefeller commission to improve medical education in China, which resulted in the founding of Peking Union Medical College. Here he is shown (front row, third from left) with John D. Rockefeller (holding his hat and immediately to the right of Welch).*

Channeling Philanthropy to Medical Science

Motivated by his desire to expand and transform medical research, Welch became a driving force in guiding the flow of philanthropic money into scientific medicine. As Barry points out in *The Great Influenza*, support for medical research by governments and universities on the Continent was one of the major reasons Europe was racing ahead of America in medical breakthroughs in the late nineteenth century.[37] At the time Johns Hopkins School of Medicine opened, Barry notes, theological schools in the United States held total endowments roughly thirty-six times larger than American medical schools.[38] Welch had observed the financial gap that affected American medicine during his two stints studying in Europe and was uniquely positioned to change it by virtue of his influence and reputation.

Welch was named as the first president of the Board of Scientific Directors of the Rockefeller Institute for Medical Research in 1901. By 1906, he was also chair of the executive committee of the Carnegie Institute. In these roles, Welch was able to direct the flow of money for medical research emanating from the two leading philanthropic organizations in the country. In this way, he helped finance the scientific revolution in American medicine that he had long envisioned.

When it came to money, Welch's engaging personality and outsized public image helped considerably. Abraham Flexner, Simon's younger brother and the author of the 1910 Carnegie report on American medical schools, was once asked, "What would you do, if you had a million dollars with which to make a start in the work of reorganizing medical education?" Flexner immediately shot back, "I should give it to Dr. Welch."[39]

Welch was about to turn sixty-seven when the United States entered World War I in 1917. He had already rendered valuable public service to his country. Nonetheless, he volunteered and was appointed a major in the medical section of the Officers' Reserve Corps (figure 3.6). Initially, his duties seem to have been evaluating the quality of medical care received by the tens of thousands of soldiers the United States would be sending to join the conflict in Europe. As one of his superior officers wrote, "Dr. Welch had a way of understanding the local situation perfectly and explaining it to the men on the ground."[40] Whether it was because of his age when he put on a uniform or out of personal idiosyncrasy, Welch never mastered military decorum. In his biography of Welch, Simon Flexner described Welch's nonchalant bearing this way: "He replied to the crisp salutes of his inferior officers by lifting his hand vaguely in the direction of his hat and gently wiggling the fingers."[41]

FIGURE 3.6. *Welch as a major in the medical section of the Officers' Reserve Corps. Here (second from right) he was photographed in 1919 with fellow medical officers on the steps of the Roman amphitheater in Arles, France.*

Confronting a Deadly Pandemic—Then and Now

Despite his lack of spit and polish, Welch played a significant role in the military's response when the influenza pandemic broke out in army barracks in the United States during the fall of 1918. As a leader of American medicine, he was one of the first people consulted when the dimensions of the contagion became apparent. Rufus Cole, the first director of Rockefeller University Hospital and a Johns Hopkins medical school graduate, wrote, "The only time I ever saw Dr. Welch really worried and disturbed was in the autumn of 1918, at Camp Devens near Boston . . .

There was a continuous line of men coming in from the various barracks [to the hospital] . . . most of them cyanosed and coughing . . . Owing to the rush and great numbers of bodies coming into the morgue, they were placed on the floor . . . we had to step amongst them."[42]

Faced with an impending pandemic, Welch's next steps have much to teach us today. Immediately after leaving Camp Devens, according to Barry, Welch made three telephone calls.[43] He asked his friend and colleague Burt Wolbach at the Brigham Hospital in Boston to autopsy patients to study the pathology of the infection. (This was crucial because, unlike today, doctors didn't know precisely how the disease damaged human tissues.) He then rang Oswald Avery, a pioneering molecular biologist at the Rockefeller Institute, asking him to research the disease. (Avery later concluded the disease was caused by a virus.) And finally, Welch urged Charles Richard, the US Army's acting surgeon general, to make "immediate provision . . . in every camp for the rapid expansion of hospital space."[44] With these three phone calls, Welch identified and set in motion the diagnosis, research, and public health processes needed to begin to fight against the pandemic. Welch himself contracted influenza, probably at Camp Devens, and became seriously ill before recovering. Nonetheless, the generation of professionals he trained and inspired, armed with the scientific approach to medicine that he promulgated, paved the path to understanding and beginning to combat a terrifying disease.

According to a Centers for Disease Control (CDC) website devoted to the history of pandemics, efforts to control the 1918 influenza pandemic were severely hampered by the lack of a known vaccine and by the complete absence of antibiotics to fight the secondary bacterial infections that often accompanied the disease. Without these prevention and treatment tools, the CDC notes, "control efforts worldwide were limited to nonpharmaceutical interventions such as isolation, quarantine, good personal hygiene, use of disinfectants, and limitations of public gatherings."[45] Several of the measures that have been tried in our own times to combat the spread of COVID-19 were known in Welch's day. Notably, wearing masks, moving routine activities outdoors, and keeping windows open were very much part of the effort to contain the 1918 influenza pandemic. Remarkably, like today, physicians even experimented with using various forms of convalescent plasma to treat infected patients.

The Public Man and the Private Self

On his eightieth birthday in 1930, Welch appeared on the cover of *Time* magazine. By then, he was the face of American medicine and had built its new foundation on the Johns Hopkins model (color plate 5). But what kind of man was William Henry Welch away from the limelight? We have alluded to the fact that his private persona was elusive—and seemingly quite different from his approachable public self. That said, we are on firmer ground in trying to understand Welch if we begin by talking about his most visible traits. He was a notorious procrastinator. Volume 1 of *The Johns Hopkins Hospital Reports* was supposed to be on pathology and volume 2 on medicine. Physician-in-Chief William Osler started work on his contribution after Welch but completed volume 2 in 1890. Welch's volume 1 wasn't published until 1897.[46]

Despite his tendency to put things off, Welch was quick to be generous. One of his students told this story: "As I started to leave the dining room, Dr. Welch said, 'Hume, I know you are going to be hard-pressed financially. I have cut out your signature from two or three of your letters to me and have pasted them on a bank deposit card at the Mercantile Trust Company. I have deposited three hundred dollars there in your name, and they will honor your signature.'"[47] Other coworkers and acquaintances testified to Welch's willingness to put his resources at their disposal, usually without being asked.

Welch had a good sense of humor and enjoyed practical jokes. At the age of eighty-two, he was traveling in Europe with neurosurgeon Harvey Cushing, a former Hopkins trainee and colleague. Cushing reported:

> At about 6:30 [A.M.] there was a knock on my door and hopping up I opened it to find [Welch] standing there in his pajamas. "There's trouble in store for us," said he. "Please come to my room immediately. There's has been a murder there . . .

> After I had turned in I was aroused by someone opening the French window. I leapt from the bed and grappled with him and finally hurled him over the balcony onto the street—not without serious loss of blood. We must get out of Paris before the police learn of this."[48]

Cushing hurried to Welch's room and opened the window. On the balcony was a bloody pile of bath towels and newspapers. Welch had spilled a bottle of red ink in the middle of the night and had decided to pull a prank on his friend.[49]

On another occasion, Welch approached his good friend William Halsted at the Maryland Club, one of their favorite haunts, after hiding a small rubber bulb, rigged to another one in his pocket, under his dress shirt. He told Halsted he had been experiencing pain in his chest after climbing the club's stairs and persuaded the famed surgeon to put his ear to his chest to determine whether he was having a heart attack or suffering from a giant pulsating aneurysm.[50]

According to one authoritative source, people who knew him well reported that Welch was a lover of cigars ("He would smoke anything that would burn").[51] He was also an avid Orioles fan, writing to his sister, "I have attended most of the games and have become an enthusiast, even a crank, on the subject."[52] Welch lived alone in rented rooms and later in a rented house on St. Paul Street in Baltimore. Most of his leisure time was spent at private clubs, notably the University Club and the Maryland Club. The University Club, in particular, was at the center of his activities. It was often two or three in the morning before Welch shut his book and wandered home. "I never go home," he said, "until after the bandits have gone to bed."[53] He was quite fond of puzzles. Welch sent his sister this missing letter puzzle with the following challenge:

> —RTHDXXFRDDNSKNWGDLDPRTFRMLGWD
> —Insert a vowel [the same vowel] in the proper places to obtain a sentence indicating knowledge gained at one of our oldest universities.
> [Answer: Orthodox Oxford dons know good old port from logwood.][54]

Welch's diversions were not all cerebral. Even in his day, Atlantic City had a slightly seedy quality. Still, he loved going there (figure 3.7). "Doubtless this is a vulgar place," he admitted, "but I am contented."[55] One roller coaster especially had a magnetic effect on him. "There is the most terrifying, miraculous, blood-curdling affair called the 'Flip-flap railroad' . . . After being pulled by trolley to the top you go down from a height of about 75 feet."[56]

FIGURE 3.7. *Welch at the beach in Atlantic City dressed in his bathing suit and smoking a cigar.*

If we try to look more deeply into Welch's private life, its heart remains mostly impenetrable. After the aching pain of ending his relationship with Frederic Dennis when Welch was in his thirties, he seemed to have walled himself off from any further intimate connection with another human being. Simon Flexner, the head of the Rockefeller Institute for Medical Research and a close collaborator, wrote that Welch had a "lifelong habit of not confiding in anyone, irrespective of who that person was or what he knew. Always he had been surrounded with people, and during most of his life he had moved on a public stage towards public ends, but always he had kept the inner core of his being inviolate."[57]

Welch never married. Indeed, he seems to have been adamantly opposed to the institution, calling it at one point early in his career "a holocaust."[58] In addition to his fondness for Atlantic City, he often visited Turkish bathhouses in Baltimore.[59] As we noted, his students privately but affectionately called Welch Popsy. They had a ditty about him that went:

> Nobody knows where Popsy eats,
> Nobody knows where Popsy sleeps,
> Nobody knows whom Popsy keeps,
> But Popsy.[60]

Looking back, it seems entirely possible that Welch was gay. In describing Welch's relationship with Frederic Dennis, a distant relative of Welch's wrote, "He liked Dr. Dennis greatly as a friend and probably as a companion."[61] Welch's likely sexual orientation would be readily accepted in most academic communities today, but would have been quite scandalous for an eminent public figure, especially one involved in health and medicine, in his era. If he was gay, the tragedy was that he never was able to express himself, never able to open up about his sexuality. This must have had a profound impact on Welch's life.

A Life to Remember

On February 1, 1933, Welch was admitted to Johns Hopkins Hospital to be treated for prostate cancer. Protocols were different then, and he apparently set up shop in the hospital for well over a year. He died at Hopkins on April 30, 1934. During his lifetime and in the wake of his death, Welch was admired—even lionized—by friends and colleagues both for his professional accomplishments and his simple humanity. His Yale roommate Lauriston Scaife wrote Welch after graduation "to express my great indebtedness for the kindness which you always manifested towards me, the pure example you set me," adding, "Billy, I feel now more deeply than ever the truth of what I have often said to others if not to you—that I was utterly unworthy [of] such a chum as yourself."[62] As we have seen, William Osler, along with Welch, one of the School of Medicine's founding physicians, had paid tribute in his diary to Welch's generosity, even temperament, and deep but amicably expressed convictions.

Welch's biographer Donald Fleming reminds us of the two seemingly opposite but complementary aspects of Welch's personality that made him uniquely suited to foment a revolution in his profession. "The secret of his career," Fleming wrote, "lay in combining a personality that breathed comfort, serenity, and ease, with a deep and overriding intellectual commitment to rebellion against the status quo in medicine."[63]

Perhaps Caroline Halsted, the Hopkins nurse who married William Halsted, said it best when she wrote to Welch after her husband's death in 1922: "I sometimes wonder if you have any idea of the great love and admiration that Dr. Halsted had for you. He never mentioned you without praise . . . I never heard him say anything but what a true friend you had always been to him and how much you and all your work meant to the Johns Hopkins. He often said, 'Welch is the Johns Hopkins. Without him it would be nothing.' 'He has been the best friend to me that any man could ever be.'"[64]

These quotations capture the spirit of William Henry Welch. They reveal what a kind individual he was, what a talented physician he became, and what an immense impact he had on those he influenced at Johns Hopkins and around the world. The revolution at Hopkins ignited by Garrett, organized by Billings, and put in practice by Welch radically transformed American medicine—in the process reinventing its conceptions of health and disease, remodeling its hospitals, and reorganizing its medical schools. It is because of Welch that American medicine was forever changed and that it will forever be viewed through the lens of science (color plate 6).

4

WILLIAM OSLER

Philosophy of Being a Physician and Patient-Centered Teaching

Sir William Osler, the Canadian physician who joined William Welch at the School of Medicine in 1889, captured the spirit of the times perfectly when he wrote, "To have lived through a revolution, to have seen a new birth of science, a new dispensation of health, reorganized medical schools, remodeled hospitals, a new outlook for humanity, is not given to every generation."[1] Osler's gratitude for living through "a revolution" in American medicine, however, belies his own extraordinary contributions to that transformation. As a clinician, a teacher, and a philosopher of medicine, Osler is generally regarded as the greatest physician North America has produced.[2] He fundamentally and forever changed the way medical students are educated, and he defined the unique values underlying what it means to be a fully capable and compassionate physician.

Osler was born on July 12, 1849, in the remote rural hamlet of Bond Head, about forty miles north of Toronto.[3] Around the time Osler reached school age, his father moved the family to the larger town of Dundas, midway between Toronto and Niagara Falls, which offered his children a better education.[4] His parents, the Reverend Featherstone Lake Osler and Ellen Free Pickton Osler, were Anglican missionaries. They had emigrated from England to Ontario twelve years earlier. Osler was the eighth of nine children in a devout family. As a boy, Willie, as he was called, was known to be athletic and jovial.[5]

The young Osler was also quite the prankster. On one occasion, he locked a gaggle of geese in the Dundas schoolhouse, and on another, he devised a smoke bomb that filled the school's classrooms with noxious vapors. After one of his pranks—it's not entirely clear which one because there were many—Osler fled, riding bareback on a local clergyman's misappropriated horse, happily shouting to his sister, "Chattie, I've got the sack!"[6] School officials eventually expelled Willie Osler and even threatened him with jail for his repeated misbehavior, but his uncle got him off. His parents decided to send Osler off to the Barrie Grammar School forty miles away, where, according to one of his biographers, his antics continued as a member of a group of rowdy students dubbed "Barrie's Bad Boys."[7] In one episode, when a local farmer advertised for a wife in the local newspaper, Osler and a friend showed up disguised as young women—Willie as a brunette and his buddy a blonde. The farmer preferred the blonde but apparently got wise before any wedding date was set. A hell-raiser for sure, Osler also revealed early on that he was creative, confident, and not afraid to take on authority, all qualities that helped him as he worked to reshape his profession.

As with so many pioneers in their fields, Osler's career was influenced partly by temperament and upbringing and partly by the opportunities of his times, which happened to be some of the most pivotal in the history of medicine. There was a compelling logic, a kind of seamless evolution, to the arc of Osler's life. Osler started as a practicing physician, caring for patients on the wards. Though he had dabbled in science as a medical student, it was in the autopsy room that he became a superb clinical scientist.[8] There he discovered the unique opportunity that studying patients who had died afforded to link clinical findings and the symptoms of disease with underlying physical changes in the organs and tissues.[9] Osler's insistence on instructing his students through direct observation of the human body—both at the bedside and in the autopsy room—made him a pathbreaking educator.[10]

As Osler's renown as a teacher and physician-scientist grew, he attracted more attention from his medical colleagues. He received invitations to join and eventually lead prestigious medical societies and, ultimately, an offer to write a textbook. Osler's landmark text, *The Principles and Practice of Medicine* (1892), highlighted the importance of the physical examination and brought his readers up to date with the revolutionary advances in fields like bacteriology and physiology that had occurred in the late 1800s. Osler also used the book to focus the medical community's attention on the lack of effective treatment for the vast majority of deadly or debilitating diseases.[11] Osler's willingness to candidly explore the limitations of medicine in his text foreshadowed his final and perhaps most far-reaching claim on our attention: his role as a philosopher of medicine.

While Osler continued to win recognition and accolades for his work as a clinician and scientist, at the pinnacle of his career, he came to be known for his reflections about what it means to be a physician. In books, articles, and scores of lectures, Osler set out to define the ethical principles and human values doctors should exemplify if medicine were to move beyond being just a learned craft and become a true profession. As we will see when we turn to Osler's philosophy, he beautifully expressed so many facets of what it means to be a physician that it is hard to briefly summarize his thinking. Nonetheless, any effort to encapsulate Osler's overarching themes must certainly include this statement in a 1903 address to students: "The practice of medicine is an art, not a trade; a calling, not a business; a calling in which your heart will be exercised equally with your head."[12] There can be little doubt that Osler approached the science of medicine as a humanist, but given the breadth and depth of his thinking about being a physician, we have tried here to let Osler speak for himself. Suffice it to say in the tens of thousands of words Osler wrote about the obligations of physicians, he never wavered from his focus on the duties that those in his profession had to work tirelessly to improve their knowledge and skills, to treat their patients not just as "cases" but as entire thinking and feeling human beings, and, finally, to behave toward their colleagues with respect and empathy.

Education and Early Career

Despite his youthful penchant for misbehavior, Osler was an able student. He started medical school at the University of Toronto and then moved to McGill Medical School, considered Canada's finest, graduating in 1872. Appointed in the same year as pathologist to the Montreal General Hospital, Osler's extraordinary dedication to his patients was displayed when he volunteered on the smallpox ward, where he contracted a mild case of the disease.[13] Like his future Hopkins colleagues Welch, Halsted, Kelly, and many other ambitious young North American physicians, Osler departed for advanced training in Europe, where he absorbed the latest medical research and techniques.

Osler returned to Montreal in 1874, first as a lecturer and then as a professor in McGill's Institutes of Medicine. He was a practicing physician seeing patients on the wards, and his drive to make these clinical cases the center of his scientific investigations was born early in his career. Osler performed over one thousand autopsies during his career and used the results to educate himself and his students.[14] He would directly compare his autopsy results with his clinical findings. As just one example, Osler found it uniquely instructive to carefully compare the sounds he heard listening to a patient's heart at the bedside to the subsequent findings when the heart was dissected.[15] In his adopted city, Osler achieved prominence as a teacher, clinician, and researcher in various posts at McGill and the Montreal General Hospital, eventually being elected president of the Canadian Medical Association in 1884.

In that same year, Osler departed Montreal to become a professor of clinical medicine at the University of Pennsylvania. A Canadian might have seemed an unlikely choice, but Osler had already been elected to the prestigious Royal College of Physicians in London and was the Montreal correspondent for the widely read Philadelphia-based *Medical News*.[16] He was well known and highly respected by prominent members of the Philadelphia medical establishment. Before Osler accepted his new post, there were protracted negotiations with his Penn counterparts and even an offer by McGill to triple his salary and create for him a chair of pathology. In the end, Penn's invitation won out, and Osler bid Montreal a bittersweet farewell, but not before Palmer Howard, his mentor at McGill, toasted him as the "one single disciple of pure science in their midst."[17]

Growing Renown

Osler stayed at the University of Pennsylvania for only five years, but his time there burnished his standing and broadened and deepened his reputation as a clinician, as a teacher, and, increasingly, as a statesman-philosopher. Decades later one of his Penn students described Osler this way:

> His first ward class was an eye-opener. In it he fairly frolicked in enthusiastic delight, and in a few moments had every man intensely interested and avid for more. Every new specimen that he came to at autopsy, and every interesting manifestation of disease in the living, was to him a treasure . . . Osler did more than any other man of his day in this city to teach all men that disease is not a horrid thing that some people, called medical men, have to study to earn a living, but a pursuit which a properly trained mind can follow with . . . keen enjoyment and uplift . . . Before Osler came the student was prone to regard cancer as a cancer; when Osler left the student studied it as an aggregation of cells possessing untold mysteries to be unravelled.[18]

In Philadelphia, Osler continued his practice of teaching students by having them move between the patient's bedside and the autopsy room to gain a more complete and definitive understanding of the physical manifestations of diseases. Osler's thinking about the power of autopsying patients whom physicians had examined clinically was perhaps based on the work of Giovanni Battista Morgagni (1682–1771), the father of anatomic pathology. In his textbook *De sedibus et causis morborum per anatomen indagates* (Of the Seats and Causes of Diseases Investigated through Anatomy) Morgagni declared, "Symptoms are the cry of suffering organs."[19] Osler was accustomed to performing autopsies at the university hospital with his students looking on (figure 4.1). He used his own money to build the hospital's first pathology lab. Another student who worked with him in the "dead house" captured Osler's way of working:

> Osler with his own and any other resident that could get away from his wards repaired to the autopsy room about two P.M. . . . The attendant would place a body on each table . . . After the thorax & abdomen were opened Osler would view the organs in situ, the residents following him from body to body. Then we would all return to the task of removing the organs & placing them on platters. Then Osler would examine the groups of organs in turn going over every one of them in the minutest detail. His joy at finding something new or unusual was quickly shared with and enjoyed by us—I can see him now with his head bent over the table suddenly exclaim, "hoity-toity, boys, look at this."[20]

Osler was not bashful about pursuing autopsies to satisfy his thirst to understand the appearance and progression of diseases. In cases where family members or friends were not present when the death occurred, consent had to be secured by tracking down the bereaved. Osler's residents knew, as one wrote, that it was "a distinct disgrace" to fail to get the required permission.[21] Osler himself was known to pursue the loved ones of deceased patients who had died at home and ask for consent to conduct an autopsy there. Even more audaciously, he would routinely request permission to take (or "borrow") a particular organ for research purposes.

FIGURE 4.1. *Osler performing an autopsy at Philadelphia General Hospital, circa 1887. He is seated, wearing a bowler hat. Osler did not wear gloves, and in fact had tuberculous lesions on his hands.*

Osler's exceptional knowledge and skills as a diagnostician didn't prevent occasional mistakes, but he was humble enough to admit them. Once, after a long day of autopsies, Osler examined a man whom he had admitted to the ward with the clinical diagnosis of pneumonia. Listening to the patient's chest, he detected a minor "dullness" but felt it was typical under the circumstances. The man died the following day. When the autopsy revealed the patient's chest was full of fluid, Osler "sent especially for all those in his ward classes, showed them what a mistake he had made, how it might have been avoided, and how careful they should be not to repeat it."[22] The outcome probably would have been the same, but the beauty of Osler's insistence on comparing diagnoses on the wards with postmortem findings was that it gave physicians a unique opportunity to learn from their mistakes.[23]

To augment his modest salary as a professor at Penn, Osler opened a private practice in one of Philadelphia's more desirable neighborhoods. His income mounted rapidly as referrals from his patients and his medical colleagues grew. According to one biographer, Osler became something of a "doctor's doctor" when many of his fellow Philadelphia physicians consulted him about their own conditions or those of relatives and friends.[24] Among Osler's famous patients were William Williams Keen Jr., one of the first brain surgeons in the United States, and Walt Whitman, the renowned poet.

His growing reputation also garnered Osler invitations to join prestigious groups such as the Philadelphia College of Physicians as well as requests to speak before important medical societies both at home and abroad. In 1885, Osler received the coveted honor of being asked to deliver the Goulstonian Lecture before the Royal College of Physicians of London.[25] His chosen topic was malignant endocarditis, an almost always fatal infection of the heart valves. Osler's discussion of the condition drew on the extensive bacteriological studies he had conducted in Montreal and was "a *tour de force* of pathological and clinical detail about a still-baffling condition."[26] A year later, Osler appeared before another demanding audience, the alumni of the College of Physicians and Surgeons in New York, and delivered a lecture titled "On Certain Problems in the Physiology of the Blood Corpuscles."[27] The talk expanded on the pioneering work he had been doing on blood platelets since his Montreal days.

Becoming a Medical Statesman

Increasingly, Osler was being asked to address professional and philosophical issues in the practice of medicine. He returned to Montreal in September 1885 to deliver his inaugural address as president of the Canadian Medical Association (CMA). In his remarks, Osler argued for raising standards and professionalizing medical education.[28] (Canada already had provincial medical boards that licensed physicians and avoided the "free trade in diplomas" that made training doctors in the United States, according to Osler, tantamount to "manslaughter.") He also advocated closing Canada's worst medical schools and spoke of the obligation of professors "to meet their brethren and give to them an account of their stewardship—for do they not hold their positions in trust?"[29]

In an interesting sidelight, perhaps with implications for today, Osler went on from his CMA address to speak at the opening ceremonies for a new building at McGill's medical school. The speech was made in the midst of Montreal's worst smallpox epidemic. We have no record of any direct comment he made on the raging infection that killed thousands of children in this French Canadian city. We do know that Osler was a steadfast advocate of vaccination throughout his career and objected to the quackery of "anti-vaccination . . . charlatanism."[30]

As a leading physician in both his native and adopted countries, Osler felt obliged to weigh in on the thorny issue of admitting women to Canadian medical schools. His views turned out to be remarkably progressive, especially compared to most of his peers. Despite opposing the creation of separate medical colleges for women, Osler wrote, "My sympathies are entirely with them in the attempt to work out the problem as to how far they can succeed in such an arduous profession as that of medicine."[31] Osler well understood the difficulties women faced even after gaining admission to medical school. That may well explain the pessimistic tone of his first conversation with Dorothy Reed when she arrived at Hopkins medical school. We learn more about that conversation when we encounter Reed in chapter 8.

South of the forty-ninth parallel, Osler's advice and influence were also being sought in debates about the future course of organized medicine. He helped to launch the Association of American Physicians (AAP). The new organization was to have one hundred members, selected by invitation and was to be devoted to advancing scientific medicine. Osler happily described the birth of the AAP, which he would later lead, as the "coming of age party of clinical medicine in America."[32] He would continue throughout his career to play a role as a leader in various medical organizations and policy debates, but as time went on, and especially after he arrived at Hopkins, he would become increasingly focused on the philosophical question of what it means to be a physician.

The Years at Hopkins

In chapter 2 we recounted how John Shaw Billings swept into Osler's residence in Philadelphia, offering him the job of chief of medicine at the nascent Johns Hopkins Hospital.[33] The groundwork for accepting Billings's offer may have been laid a few years earlier when shortly after arriving in Philadelphia, Osler paid a visit to Johns Hopkins University. He wrote a friend in Ontario, "It is the university of the future & when the Medical School is organized all others will be distanced in the country."[34] Osler's arrival in Baltimore in 1889 was a turning point not only for the practice of medicine but also for the development of its philosophical foundations.

As physician-in-chief at a completely new institution, Osler was given carte blanche in organizing the medical staff. Modeling the organization of faculty and trainees on what he had observed in Europe, Osler established a residency program at Hopkins and assembled a handpicked team of experienced resident physicians, some of whom remained under his supervision for five years or more—very different from the loose hierarchy in most American hospitals. Osler's first chief resident was Henri Lafleur, a fellow McGill graduate whom Osler had recruited.[35] Lafleur and the other residents took over the day-to-day care of the patients, allowing Osler to focus on teaching and writing his book. As a clinical teacher, Osler, according to one of his students, was "delightfully informal, erudite beyond comparison, entertaining but surprisingly effective."[36] By far the most important accomplishment during Osler's first years in Baltimore was his groundbreaking textbook *The Principles and Practice of Medicine*.

Osler had been thinking about writing a textbook for some time but had put off the grind for other more appealing projects. In early 1891, with a reinvigorating summer trip to European medical capitals behind him and no signs that the School of Medicine would open anytime soon, he signed a contract with a New York publisher.[37] Osler later called the project "selling my brains to the devil,"[38] but the impact of his textbook cannot be overstated.

As Osler's biographer Michael Bliss notes, *Principles and Practice* was "at once a monument to the achievements of nineteenth-century scientific medicine and a gateway to the twentieth century."[39] Osler demonstrated a firm command of the germ theory breakthroughs of the 1880s that had cleared a path to understanding the causes of many infectious diseases, even if antibiotics to treat them had yet to be discovered. Diagnostic tools, including X-rays and electrocardiograms, were still in the offing, but Osler used the knowledge he gained autopsying patients he had examined and treated to create unique insights into the clinical manifestations of disease.[40]

The Principles and Practice of Medicine was essentially a thousand-page-plus compendium of illnesses, their symptoms, and their treatments or—as Osler was keenly aware—the absence of effective therapies for most conditions. The lack of useful treatments for most debilitating or deadly diseases was a grave concern for Osler. He sometimes prescribed medications to alleviate pain and suffering and also because he believed that is what patients wanted their doctors to do. Osler was quick to prescribe opium for relatively minor ailments such as "the distressing, irritative cough, which keeps the patient awake," but he was generally skeptical of over-drugging and especially of the pharmacological "cocktails," composed of various combinations of ineffective ingredients, that were becoming popular in his day.[41] By and large, Osler's philosophy was not to overprescribe. In fact, the frustration Osler expressed in his textbook with a lack of effective therapies led Reverend Frederick T. Gates, a close adviser to John D. Rockefeller, to encourage Rockefeller to focus his philanthropy on medicine.[42]

Ultimately, Osler was a realist and understood the limitations of current medical knowledge when it came to remedies for bacterial infections and many other conditions. Even with his encyclopedic knowledge and experience, he was humble. He opened his textbook with a quotation from Hippocrates: "Experience is fallacious and judgment difficult."[43]

The success of *Principles and Practice* reflected both the receptive market for medical textbooks at the time Osler wrote the book and his unique skills as an author. Most of the other widely used medical textbooks had become outdated, mainly by scientific advances, especially in bacteriology. In some cases, the authors of these texts had died, making updated editions impossible or at least unlikely. Osler's renown as a clinician and published researcher in three countries—the United States, Canada, and Britain added to the enthusiasm for his book.

Osler's clarity and economical prose also distinguished his text.[44] His habit of dictating the book to his stenographer, B. O. Humpton ("Miss Hump"), as he routinely did with his autopsies and patient notes, led to "simpler, more direct sentence structure."[45] On the other hand, *Principles and Practice* didn't skim lightly over the surface. For virtually every condition, Osler traced the full evolution of the latest medical thinking, with all of its controversies and conflicting opinions, including his own. Osler routinely used "I," that forbidden word in textbooks, and his text was engaging, filled, as renowned British physician and scientist Sir Arthur MacNalty wrote, "with personal references to his vast clinical experience, and attractive allusions to personalities, ancient and modern, in illustrations of symptoms and treatment."[46] Best of all, Osler was forthright and quick to concede the limitations of physicians' knowledge.

To write large parts of his textbook, Osler commandeered the sitting room of Hunter Robb, a resident at Hopkins Hospital, for the better part of six months (figure 4.2). Robb recalled Osler breaking off his work and suddenly rushing into the adjoining bedroom to swap stories or challenge him to a coin game. He claimed to have enjoyed Osler's interruptions, "except when he would court inspiration by kicking my waste-paper basket about the room," a problem Robb fixed by putting bricks in the receptacle.[47]

FIGURE 4.2. *Osler writing* The Principles and Practice of Medicine, *circa 1891. He commandeered the sitting room of one of the residents on the staff of Hopkins Hospital to write his textbook.*

The Principles and Practice of Medicine was an almost immediate commercial and critical hit when it appeared in 1892. The publisher, D. Appleton and Company in New York, sold out the first printing of three thousand copies in three months.[48] Osler netted $10,000 in royalties when two-year sales of the textbook reached fourteen thousand copies. It went through many revised editions and became an important part of Osler's income. Later on, Osler liked to call his accumulating fund balance at Appleton his "boodle account."[49]

Osler's Philosophy on Being a Physician

Despite his acclaimed textbook and his stellar accomplishments as a clinician and researcher, we might not remember Osler, now gone for more than a century, were it not for his philosophy. He was an extraordinary thinker on what it means to be a physician and, perhaps more to the point, on what it means to live our lives in a meaningful way. We think it's worth sharing some of his insights, quoting him directly to capture his inspiring ideals.

Bedside Teaching

Since his days at the Montreal General Hospital, Osler had taken students directly to the wards for instruction. This practice continued at Penn and grew at Hopkins. Osler had observed this teaching method in Europe but was the first to spread the message in the United States. We can picture him arriving precisely at 9:00 A.M. at the "Dome," the hospital's main entrance. His chief resident would be waiting and would take him, arm-in-arm, up to the wards, where crowds of students and trainees were gathered to watch him examine patients and discuss their cases (figure 4.3). In 1901, Osler told the Society of Internal Medicine:

> I am firmly convinced that the best book in medicine is the book of Nature, as writ large in the bodies of men. You remember the answer of the immortal Hunter [the famous

British surgeon], when asked what books the student should read in anatomy—he opened the door of the dissecting-room and pointed to the tables.[50]

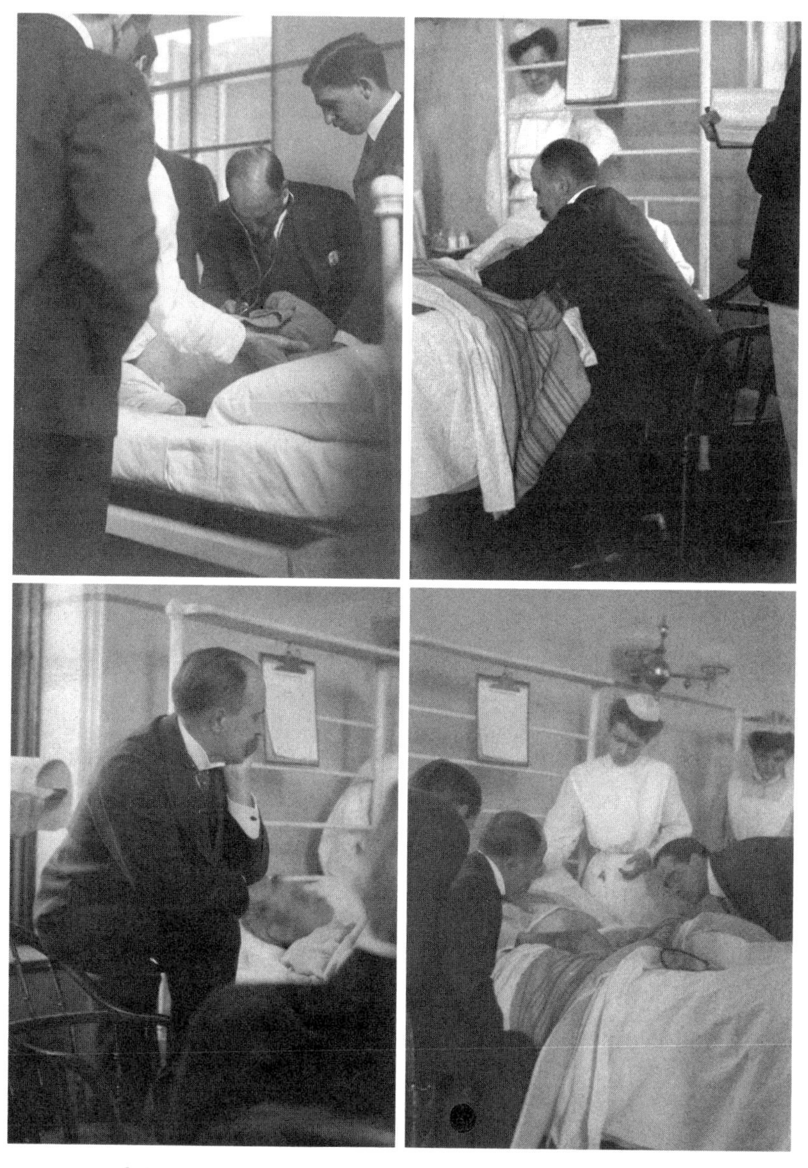

FIGURE 4.3. *Osler examining a patient. He routinely used four techniques (clockwise from upper left): auscultation (listening), palpation (touch), observation, and contemplation.*

This philosophy of seeing and doing, not just reading, was central to Osler's thinking about medical education. He declared that he wanted for his epitaph "Here lies the man who admitted students to the wards"[51] (figure 4.4).

FIGURE 4.4. *Osler desired no epitaph other than that he taught students on the ward. On or off the ward, his teaching was patient-centric. Here he is shown teaching students at the side of a patient's bed in the amphitheater of Hopkins Hospital, circa 1900.*

Today, in an era when patients are often diagnosed and treated piecemeal by armies of specialists connected only by the complex web of electronic medical records, we would do well to remember Osler's advice to "listen to the patient, he is telling you the diagnosis."[52] There is great value in having one's hands, eyes, and especially ears attuned to patients' own descriptions of their illness.

Imperturbability and Equanimity

Osler understood that medicine was an inherently stressful profession, involving as it does human lives and human emotions, often in the most demanding and adverse situations. He believed that to succeed physicians must have two qualities. The first was imperturbability, an almost physical or instinctual rock steadiness:

> Imperturbability means coolness and presence of mind under all circumstances, calmness amid storm, clearness of judgement in moments of grave peril, immobility, impassiveness. (From Osler's valedictory address "Aequanimitas" at the University of Pennsylvania, May 1, 1889)[53]

The second quality doctors needed was equanimity, the mental equivalent to the bodily endowment of cucumber-like coolness. Osler recognized that some human beings are born with imperturbability; others are not. But he insisted that every physician could and should strive for equanimity:

> Cultivate, then, gentlemen, such a judicious measure of obtuseness as will enable you to meet the exigencies of practice with firmness and courage, without, at the same time, hardening "the human heart by which we live." (From Osler's valedictory address "Aequanimitas")[54]

Osler's philosophy lives on at Hopkins. The scarves and ties that the residents in the Department of Medicine at Johns Hopkins Hospital wear today have an emblem on them that reads "Aequanimitas," a reminder of Osler's injunction to stay even-tempered.

Work: The Secret of Life

This tenet of Osler's philosophy needs little explanation and may at first seem trite. He believed that work was the secret to a successful and fulfilled life. In this regard, Osler absolutely walked the talk. His clinical and scholarly output was prodigious. The following lines are among the most often quoted of Osler's many quotable quotations.

> I propose to tell you the secret of life as I have seen the game played, and as I have tried to play it myself . . . Though a little one, the master-word looms large in meaning. It is the open sesame to every portal, the great equalizer in the world, the true philosopher's stone, which transmutes all the base metal of humanity into gold. The stupid man among you it will make bright, the bright man brilliant, and the brilliant student steady . . . It is directly responsible for all advances in medicine during the past twenty-five centuries . . . And the master-word is Work, a little one, as I have said, but fraught with momentous sequences if you can but write it on the tablets of your hearts. (From a speech given in 1903 at ceremonies celebrating the merger of Trinity College and its medical school with the University of Toronto)[55]

Today, in an era in which students seem eager to skip steps, Osler's words remind us that there is no substitute for hard work.

Consume Your Own Smoke

Osler had his own version of "don't sweat the small stuff." He cautioned us not to waste our energies and lose our focus complaining about life's minor aggravations.

> Learn to consume your own smoke. The atmosphere is darkened by the murmurings and whimperings of men and women over the non-essentials, the trifles, that are inevitably incident to the hurly-burly of the day's routine. Things cannot always go your way. Learn to accept in silence the minor aggravations, cultivate the gift of taciturnity and consume your own smoke with an extra draught of hard work, so that those about you may not be annoyed with the dust and soot of your complaints. (From Osler's 1903 speech celebrating the Trinity College–University of Toronto merger.)[56]

How many of us have seen perfectly capable coworkers get caught up in petty aggravations that poison the work environment?

Day-Tight Compartments

The famed British historian and essayist Thomas Carlyle wrote, "Our main business is not to see what lies dimly at a distance, but to do what clearly lies at hand." Osler frequently drew from Carlyle, as he did when he wrote about the wisdom of managing time by compartmentalizing. Easier said than done in today's anxiety-prone society, but Osler was a firm advocate of ignoring tomorrow's cares and discarding yesterday's mistakes to face the challenges of today.

> The load of tomorrow, added to that of yesterday, carried today makes the strongest falter. Shut off the future as tightly as the past . . . who can tell what a day may bring forth? . . . The future is today—there is no tomorrow! . . . The life of the present, of today, lived earnestly, intently, without a forward-looking thought, is the only insurance for the future. Let the limit of your horizon be a twenty-four hour circle. (From "A Way of Life," a speech given to students at Yale in 1913)[57]

Here Osler urges his listeners to divide and conquer their overly long lists of worries and focus only on the urgent tasks that confront them today. For all of us, the weight of worrying about yesterday's mistakes or tomorrow's insurmountable hurdles only heightens our anxiety and does nothing to help us accomplish the task at hand.

Lifelong Learning

Osler was absolutely a believer in the notion that learning never ceases (figure 4.5). He urged physicians to spend the last thirty minutes of the day "in communion with the saints of humanity."[58] His own bedside library included, among other classics, the Bible, Shakespeare, *Don Quixote*, and

Emerson. Today, many of us, truth be told, would have to admit spending our last waking half hour communing with our cell phones!

> The great fundamental principle laid down by Plato—that education is a life-long process, in which the student can only make a beginning . . . To cover the vast field of medicine in four years is an impossible task. We can only instill principles, put the student on the right path, give him methods, teach him how to study, and early to discern between essentials and non-essentials. (From "After Twenty-Five Years," an address to the faculty and students at McGill, 1899)[59]

FIGURE 4.5. *Osler reading at his home in Oxford. He emphasized the value of lifelong learning and was a tireless reader.*

The classic literature by Osler's bedside is not just a quaint reminder of a bygone era. The classics are classics precisely because they provide insights into human nature and mortality that remain as relevant to being a physician today as when they were written.

Respect for Other Physicians

Osler knew that, in a profession where human emotions and personal judgment play such a strong part, the temptation to speak harshly of fellow physicians is strong. He cautioned against it.

> Many a physician whose daily work is a daily round of beneficence will say hard things and think hard thoughts of a colleague. No sin will so easily beset you as uncharitableness towards your brother practitioner. So strong is the personal element in the practice of medicine, and so many are the wagging tongues in every parish, that evil-speaking, lying and slandering find a shining mark in the lapses and mistakes which are inevitable . . . never under any circumstances listen to a tale told to the detriment of a brother practitioner. (From "The Master Word in Medicine," an address at the University of Toronto, 1903)[60]

Osler's admonition to avoid listening to or relaying malicious gossip is perhaps more important today, as we face the constant buzz of emails, texts, and social media.

Love of Humanity

At the core of Osler's philosophy of medicine, and of life, was his deep empathy for other human beings. He was able to see the good in people not only in his own circle but in most people he encountered.

> Nothing will sustain you more potently than the power to recognize in your humdrum routine, as perhaps it may be

thought, the true poetry of life—the poetry of the commonplace, of the ordinary man, of the plain, toil-worn woman, with their loves and their joys, their sorrows and their griefs. (From "The Student Life," a farewell speech given before Osler moved to Oxford University in 1905)[61]

FIGURE 4.6. *Osler loved children. Here he is seen playing with his son Revere.*

Osler often inscribed on gifts he gave to others these words from the poem "Abou Ben Adham" by Leigh Hunt: "Write me as one that loves his fellow men."[62] Osler's example of deep love and faith in humanity lights our path.

Predictably for someone who held an optimistic view of humanity, Osler loved children (figure 4.6). His biographer Michael Bliss describes him as having had a magical effect on them. Bliss notes that Osler was the kind of person "who would have stood out, did stand out, in any crowd—except that he had a habit of disappearing from crowds to find some children to play with."[63]

Marriage, Family, and Departure for England

Turning back from Osler's philosophy to his life story, we find that shortly after completing his seminal textbook he married Grace Revere, the great-granddaughter of Paul Revere and the widow of Samuel W. Gross, the famous Philadelphia surgeon. While at Penn, Osler had befriended Gross, who had succeeded his equally illustrious father as chair of surgery at Jefferson Medical College. Osler was forty-two and Grace Revere thirty-seven when they wed in May 1892.[64] The couple had two children. Their first, a boy, was delivered by Osler's colleague Howard Kelly, chief of gynecology and obstetrics at Hopkins, but died after six days. "One must take the rough with the smooth,"[65] Osler wrote mournfully. Their second child, Edward Revere, who went by Revere, was born in 1895.

Osler's career continued on its successful path at Hopkins, which was breaking new ground in medicine and medical education in the 1890s and early 1900s. He published important papers and delivered pathbreaking lectures on conditions ranging from erythema multiforme (a particular rash that often starts on the hands and feet) to cerebrospinal fever (meningitis) and chronic cyanosis (a condition where the skin is blue from lack of oxygen). In 1895 he was named president of both the Association of American Physicians, which he had helped found, and the Association of American Medical Colleges.[66] He was named dean of the Johns Hopkins Medical School in 1898.[67] All the while, Osler continued to collect fellowships and honorary degrees from medical societies and universities from Aberdeen to Yale. The scholarly work, leadership responsibilities, and ceremonial duties were on top of his mounting clinical and teaching duties.

People across the country and around the world were coming to Baltimore to be treated by Osler. His patients included the wealthy and the poor, the prominent and the not-so-prominent. Finally, the stress of so many people depending on him became too great. Osler wrote, "The racket of my present life is too much for me. I am going downhill physically & mentally."[68] In 1905 Osler accepted the position of Regius Professor of Medicine at Oxford University. The appointment appealed to Osler in part because Oxford didn't have an affiliated hospital, which would leave him at peace and with more time for his writing.

Although one of the attractions of Oxford was a reduction in his clinical duties, Osler hardly abandoned them. In Oxford for only months, he launched Sunday morning clinics for postgraduate students at the Radcliffe Infirmary. Even as his attention began to shift to more literary and philosophical concerns, Osler continued to produce important scientific papers, including one on angina, or chest pain, published in the *Lancet* in 1910.[69]

For all the extraordinary achievements in the final chapter of Osler's life, the mischievous boy was still within him. In 1911 he was made a baronet by the British crown (hence his title Sir William Osler). With his habitual modesty, Osler at first wanted to decline the honor, but in the end accepted it, writing his sister Chattie, "It is wonderful how a bad boy may fool his fellows if he once gets to work."[70]

While the move to Oxford brought William and Grace Osler a measure of relief from the unrelenting pressures of being chief of medicine at Hopkins, there was also a tragic consequence. Shortly after the outbreak of World War I, their son Revere, now a young man, dropped out of his studies at Oxford and volunteered with the Canadian Expeditionary Force (figure 4.7). His battalion took part in the Third Battle of Ypres in Belgium on August 29, 1917. He was hit by a German artillery barrage and badly injured. The young Osler was evacuated to the casualty clearing station, where, by chance, his care was entrusted to Harvey Cushing, on leave from his faculty position in surgery at Harvard. Cushing was extremely close to Osler and tried desperately to save the son of his mentor and friend. Sadly,

Revere's injuries were fatal. "We saw him buried," Cushing wrote, "in the early morning. A soggy Flanders field beside a little oak . . . an overcast, windy, autumnal day—the long rows of simple wooden crosses—the new ditches half full of water."[71]

FIGURE 4.7. *Osler with his son Revere in 1916 on the terrace of their home in Oxford. Revere would be mortally wounded by a German artillery barrage in the fighting at Ypres in Belgium in August 1917.*

Back in Oxford, the Oslers were left to deal with the grief of having lost their only remaining child. Though it was not in Osler's nature to publicly burden others with his loss, we do know that he had already wrestled with the problem of deep unhappiness from a philosophical perspective. In his speech "Aequanimitas" (1889), Osler had written:

> It is sad to think that, for some of you, there is in store disappointment, perhaps failure. You cannot hope, of course, to escape . . . Stand up bravely, even against the worst . . . for in persistency lies victory . . . there is a struggle with defeat which some of you will have to bear, and it will be well for you in that day to have cultivated a cheerful equanimity.[72]

Osler seems to have adopted his own prescription. After Revere's tragic death, he remained outwardly positive and optimistic, but those who knew him best recognized a fundamental inner sadness from which he never recovered. Shortly before his death, Osler wrote, "The Fates do not allow the good fortune that has followed me to go with me to the grave—call no man happy till he dies."[73]

Osler's Death and His Philosophy about Dying

Osler was an extraordinary clinician who focused on bedside teaching, but he was also a humanist who anticipated today's more compassionate and holistic views of illness, death, and dying. His humanistic thinking was exemplified in a study he undertook at the age of fifty-five observing 486 dying patients at Johns Hopkins Hospital.[74] His purpose was to understand the "sensations of the dying."[75] He concluded that "the majority gave no sign one way or the other; like their birth, their death was a sleep and a forgetting."[76] Osler clearly recognized that the modern way of dying in hospitals deprived terminally ill patients of the tender care of loved ones who knew and respected the terminally ill person's wishes. He was skeptical and generally critical of healthcare professionals who insisted on ultimately futile therapeutic measures that deprived the gravely ill patient of the "deep-seated animal instinct to . . . die undisturbed."[77]

With dying patients he cared for personally, Osler's approach was more likely to emphasize compassion and soothing the patient's fears and anxieties than last-ditch medical measures. This attitude is beautifully illustrated in the case of a small girl named Janet who was dying during the influenza pandemic that ravaged Europe in 1918, the year after Osler lost his son in the war. Janet's mother described how Osler paid his last visit to their home on a blustery November day, bringing with him a single red rose from his garden and telling Janet that, as he passed by this last rose of the summer, the flower had spoken to him. The flower insisted "that she wished to go along with him to see his little lassie." Her mother recalled:

> That evening we all had a fairy teaparty, at a tiny table by the bed, Sir William talking to the rose, his little lassie, and her mother in a most exquisite way . . . all crouched down on his heels; and the little girl understood that neither fairies nor people could always have the color of a red rose in their cheeks, or stay as long as they wanted to in one place, but that they nevertheless would be very happy in another home and must not let the people they left behind, particularly their parents, feel badly about it: and the little girl understood and was not unhappy.[78]

Whether we are caring for terminally ill patients as healthcare professionals, family members, or friends, all of us have much to learn from Osler's words to Janet.

Osler faced his own death with the same combination of sensible medical treatment to alleviate pain and calm resignation in the face of the inevitable with which he eased his own patients to their final breath. Osler was already afflicted with chronic bronchitis when he departed Hopkins for Oxford in 1905. He maintained his full round of teaching and writing activities, but his condition worsened. In October 1919, Osler began to be wracked by long and repeated fits of coughing.

By now bedridden, he requested opiates to relieve the "paroxysms" that exhausted him. They were effective, as he knew they would be. "Shunt the whole pharmacopoeia, except opium. It alone in some form does the job."[79] Books were his other comfort. When, near the end, Osler was unable to read, he asked his nephew to recite aloud his favorite poetry, including verses from Samuel Taylor Coleridge's "The Rime of the Ancient Mariner" on the day before he passed away.

Despite some discomfort from a lung abscess on his last day, apparently eased by self-prescribed opiates, Osler died peacefully in his own bed at 4:30 P.M. on December 29, 1919 (color plate 7).[80]

When we remember Osler and attempt to capture his lasting contributions, what should we think about? Certainly, we should recall his prodigious work ethic and fundamental decency towards his patients, colleagues, and friends, two qualities he considered closely linked. In his 1891 address "Doctor and Nurse," Osler told his audience, "There should be for each of you a busy, useful, and happy life; more you cannot expect . . . .We are here to add what we can *to*, not to get what we can *from*, life."[81]

We should also recall those twin North Stars of his career—patient-centered teaching at the bedside and the philosophy of being a physician. These guideposts were again complementary, blending his drive for a scientific understanding of illness with the imperative to treat those who were suffering compassionately. His achievements in both areas were historic. It is easy to make Osler into a saint of medicine, as some have done, but like all ten men and women we have profiled, he was exemplary but far from perfect. Newer scholarship has revealed the degree to which Osler held prejudiced views on race.

It is difficult, and in many ways disheartening, to make a case that someone like Osler, whose humanizing legacy in medicine has been so powerful and lasting, could be racially biased. Yet, as the authors of a recent article in the *Canadian Medical Association Journal* argue, the evidence is clear and abundant.[82] Musing about whether to attend the 1893 Pan-American Medical Congress, Osler confided in a letter to Dr. H. V. Ogden "I hate Latin Americans."[83] Even his spirited defense of the young Dorothy Reed, ridiculed by her fellow interns for losing six patients to pneumonia overnight on a ward for Black patients at Johns Hopkins, betrayed prejudice. Explaining her misfortune, Osler allowed that "the coloured, usually both syphilitic and alcoholic, were the worst risk in pulmonary disease."[84] His judgments about Native Americans could be equally harsh. An unpublished manuscript dealing with Indigenous people offered this choice racial slur: "Every primitive tribe retains some vile animal habit not yet illuminated in the upward march of the [human] race."[85] In one of Osler's last public speeches, he weighed in on the debate over a wave of Asian immigration to Canada declaring, "We are bound to make our country a White man's country."[86]

We would be remiss if we didn't acknowledge the argument frequently offered by Osler's defenders that his views on race were simply "the product of his times." At same time, we are duty-bound to offer the facts as they have been reliably recorded. Even acknowledging the passage of more than a century, these facts force us to consider Osler, sometimes presented as a paragon of medical virtue, in a more human light. Quite possibly, the most fitting ethical standard by which to measure Osler would be his own—a standard inspired by his better angels, which frequently but not always prevailed in his thinking about human affairs. Addressing the Canadian Medical Association in 1902, Osler decried nationalism and other parochial forms of intolerance, which he called "a vice of the blood, . . . it runs riot in the race, and rages today as of yore despite the precepts of religion and the practice of democracy . . . What I inveigh against is a curse and spirit of intolerance, conceived in distrust and bred in ignorance, . . . that makes the mental attitude . . . [forget] the higher claims of human brotherhood."[87]

Osler, as is true of all of us, on occasion fell short of the mark he had set for himself when it came to the love of humanity, especially those humans of different races and nationalities. Nonetheless, we have tried to balance the complex calculus of evaluating another human life. (It is a daunting task that Osler himself described as we note in our preface.) We believe it is important to consider Osler's racial biases in the context of his entire impact on medicine. From that perspective, his revolutionary view of empirically based bedside medical education and compassionate patient-centered care helped bring to life the medical school Mary Elizabeth Garrett and John Shaw Billings envisioned and transformed Hopkins into a model of its kind.

5

WILLIAM STEWART HALSTED

Victory Wrought from Defeat

We recall that Mary Elizabeth Garrett, in one of the final chapters of her long, often contentious, but ultimately transformative relationship with Johns Hopkins, presented the university with *The Four Doctors*, the magnificent group portrait by John Singer Sargent. The painting depicts William Henry Welch, William Stewart Halsted, William Osler, and Howard Atwood Kelly, the members of the founding faculty of the new school of medicine Garrett most admired, in full academic regalia (color plate 2). Art historians and generations of viewers alike have long been captivated by this evocative image of men of science dedicated to the art of healing.

Yet, even casual observers are frequently struck by one curious aspect of the famous painting: Halsted, despite being the only figure in the portrait who is standing, appears to be almost fading into the background.[1] It's as if the man widely regarded as the most innovative surgeon America ever produced wanted to remain in the shadows, away from the light that illuminates his colleagues' faces. And, indeed, he did. After a series of brilliant experiments that demonstrated the value of cocaine as a local and regional anesthetic, Halsted became addicted—first to cocaine and later, in a futile effort to overcome his dependence, to morphine. In this chapter, we will learn about the tremendous price Halsted paid to advance safe, painless surgery and about his many other extraordinary accomplishments as a pioneering surgeon in the face of this crushing burden.

To fully appreciate how Halsted revolutionized surgery, we need only look at a second painting famous in the history of American medicine. In color plate 8, we see Thomas Eakins's 1875 canvas *The Gross Clinic*, which gives us a front row seat at an operation performed by one of America's most renowned surgeons, Dr. Samuel D. Gross of Philadelphia, the year after Halsted entered medical school. The painting leaves us with a vivid sense of the sheer terror of surgery in the late nineteenth century.[2] Blood covers Gross's ungloved hand, his assistant appears to be holding down the patient, and the boy's mother covers her eyes and shrinks in terror in the background.

When Halsted started his career in medicine, surgery was a hurried race against the clock, marked by crude cuts, inattention to blood loss, and rough handling of tissues—all of it conducted under filthy conditions. Even though general anesthesia had been developed in the mid-nineteenth century, most surgeons persisted in operating on patients as fast as possible. As one surgeon described it, "Operating rooms of that day were nightmares . . . interns often held ligatures in their mouths, and little care was taken in cleaning instruments. Surgeons and nurses frequently went from operating room to operating room wearing contaminated garments, not bothering to wash hands."[3]

A New Model of Surgery

Compare Eakins's frightening portrayal of surgery with figure 5.1, which shows Halsted operating at Johns Hopkins in 1904. We can readily understand how "the father of American surgery" transformed his profession. The well-ordered scene suggests the deliberate pace, precision, and beauty of surgery conducted under aseptic conditions, with careful attention to controlling blood loss, and the delicate handling of tissues. As Halsted described it, "Pain, hemorrhage, infection, the three great evils which had always embittered the practice of surgery and checked its progress, were, in a moment, in a quarter of a century (1846–1873) robbed of their terrors."[4]

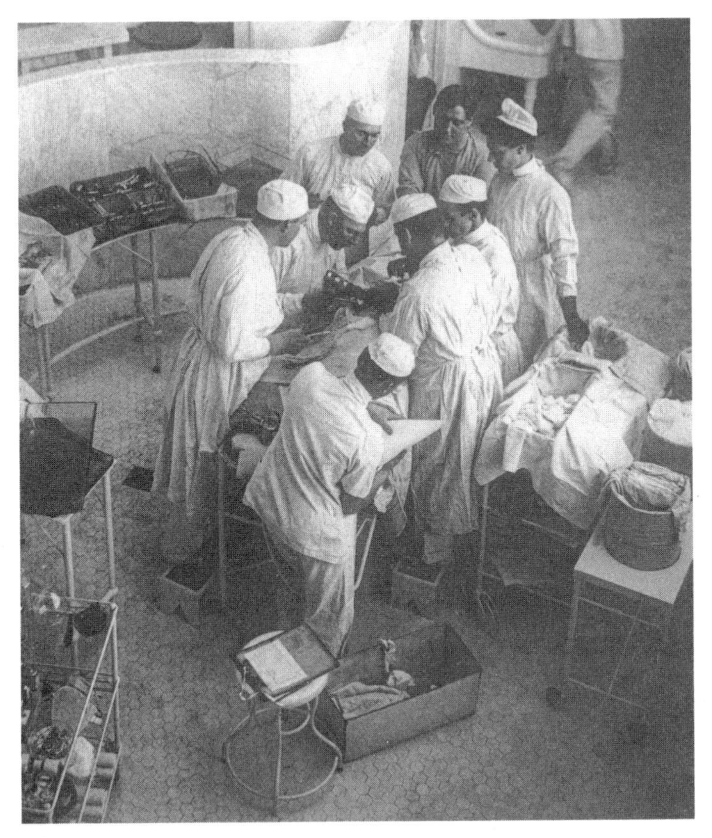

FIGURE 5.1. *Halsted (seen leaning over the patient at 11 o'clock) operating with gloves in 1904. The patient had an infection of his femur. While the surgeons and scrub nurse wore gloves, surgical masks had not yet been introduced.*

A Life Story in Two Parts

Halsted's record as a pioneer in American surgery is one of unbroken accomplishment, but the trajectory of his career has an arc with two fundamentally distinct phases. There were the early years in New York, when a young Halsted, recently graduated from medical school and just returned from Europe and his studies with some of the continent's greatest physicians, rose in prominence and established himself as a bold and inventive surgeon. Besides having a packed surgical schedule and operating regularly in at least six of the city's busiest hospitals, Halsted maintained a high level of physical activity and a busy social life. Friends from this period described him as a "model of muscular strength and vigor, full of enthusiasm and the joy of life," a "bold, daring and original surgeon," and having a "gay and cheerful disposition."[5]

Contrast these words with this appraisal of Halsted after he moved to Baltimore to take up his post at Johns Hopkins written by his surgical colleague and former trainee Harvey Cushing:

> A man of unique personality, shy, something of a recluse, fastidious in his tastes and in his friendships, an aristocrat in his breeding, scholarly in his habits, the victim for many years of indifferent health . . . Over-modest about his work . . . caring little for the gregarious gatherings of medical men, unassuming, having little interest in private practice.[6]

The turning point in Halsted's career came with his addiction to cocaine following his breakthrough experiments in the mid-1880s that established that the narcotic could be used as an effective surgical anesthetic. To understand how Halsted's astonishing productivity as a surgeon, researcher, and teacher continued even in the face of his struggle with addiction, we need to turn back to the beginning of his story.

Halsted's Early Years

William Stewart Halsted was born into a wealthy family in New York City on September 23, 1852. He was the oldest of four children born to Mary

Louisa Haines and William Mills Halsted Jr. The family's import business, Halsted, Haines & Co., had flourished, and Halsted's early life was one of privilege. In addition to their elegant Fifth Avenue mansion, the family owned a summer retreat at Irvington in the scenic Hudson River Valley. The young William's early schooling was at home with private tutors before he was sent to the Monson Academy in Massachusetts. The experience proved so unhappy that Halsted ran away from the school. His parents next sent him to the elite Phillips Andover Academy in Andover, Massachusetts. Deemed not quite ready for college after six years at Andover, Halsted spent one more year at home with private tutors and then entered Yale in 1870.

Halsted's college roommate, Samuel Bushnell, described him as "exceptionally strong, careful in matters of dress, generally popular . . . possessing a ready wit and caustic tongue, but showing no evidence of unusual abilities or great ambitions."[7] By his own description, Halsted devoted himself "solely to athletics in college."[8] Biographers can find no evidence that he ever checked a book out of the Yale Library during his four years in New Haven.[9] Halsted accomplished far more on the playing field than in the classroom. He was the shortstop on Yale's baseball team, he rowed for Yale's crew, and he was an outstanding football player (figure 5.2). In his senior year, Halsted captained the Yale team that played the first game with eleven men on each side. He scored the winning points in what was the forerunner of today's Super Bowl. Halsted also led an active social life and belonged to several of Yale's social clubs and societies. The picture that emerges from his college years is of a young man who led a charmed life on campus.

As we have seen, American medical schools had few, if any, admission standards in this era. It was relatively easy for Halsted to gain entrance to the College of Physicians and Surgeons in New York (now Columbia University's medical school) after graduating from Yale in 1874. A flavor of the state of medical education at the time comes through in this description by a medical historian of a lecture hall at the College of Physicians and Surgeons: "[It was] dimly lit, smoke and dust filled, and stifling hot, so that students not infrequently fainted. The professor's entrance was generally greeted with cat-calls, whistles, and yells, and general pandemonium reigned between lectures."[10] Despite the raucous classroom environment, Halsted's focus on his studies changed dramatically in medical school, and he buckled down to his courses. We can date his unparalleled record of accomplishment as a surgeon from this period.

FIGURE 5.2. *Yale's 1874 baseball team. A muscular Halsted stands at the far right.*

A year shy of completing the three-year program at the College of Physicians and Surgeons, Halsted applied for an internship at Bellevue Hospital. Though technically not eligible without his medical degree, Halsted did well enough on the examination to secure the position. As an intern, Halsted spent most of his time in the medical wards but also assisted in some operations. During Halsted's year at Bellevue, the British surgeon Joseph Lister visited the United States to preach the gospel of antiseptic surgery. Though many of his medical colleagues in America were reluctant to embrace Lister's ideas (as we have seen in the case of the famous Philadelphia surgeon Samuel D. Gross), Halsted clearly was a believer in the superior results obtained when surgeons worked to reduce infections. Asepsis, or the practice of keeping environments free of pathogens, became a hallmark of Halsted's meticulous surgical technique.

Halsted took the exam for his medical degree near the end of his internship and finished at or near the top of his class. Upon graduation in 1877, he was hired as the house physician at Bellevue. It was there that Halsted developed the bedside chart that tracks a patient's vital signs by recording pulse, temperature, and respiration. Today, the chart—or an electronic version—is used in virtually every hospital in the world. While at Bellevue, Halsted likely met William Welch (at that time a pathologist there), who became one of Halsted's closest friends and played a crucial role in bringing him to Hopkins.

With his two years at Bellevue, one as house physician and another as an intern, Halsted had completed all of the postgraduate medical education available to him in the United States. As Welch and Osler and many other ambitious young American physicians had done, Halsted departed for Europe to train with the most eminent physicians of his day and absorb their brand of research-driven medicine. In Vienna, he studied under Hans Chiari in pathology, Leopold Schneck in embryology, and in surgery with Anton Wölfler and the legendary Theodor Billroth. In Würzburg, Halsted dove into embryology with Albert von Kölliker, histology with Philipp Stöher, and surgery with Ernst von Bergmann. The surgeons Richard von Volkmann in Halle, Max Schede in Hamburg, and Friedrich von Esmarch in Kiel also took Halsted under their wings.[11] Less important than the famous doctors' names and reputations was the breadth and depth of advanced technical knowledge and the belief in science-based medicine that Halsted absorbed in Europe.

The Years in New York: A Bold and Daring Surgeon

Halsted returned to New York in the fall of 1880. It was an almost perfect time to launch his surgical career. As John Cameron points out in his biographical article, in the wake of Lister's introduction of the new concept of antisepsis to reduce the risk of infection, German surgeons had perfected instruments to control bleeding.[12] The two foundations of modern surgery were now in place. There had also been rapid advances in fields like bacteriology and physiology, as well as greater appreciation of the benefits of general anesthesia in managing pain. Halsted was in the right place at the right time, and he seized the opportunity.

Halsted's surgical skills were soon in demand across the city, and he worked tirelessly. He joined the staff of Roosevelt Hospital. His work to establish an outpatient clinic there was so successful that the hospital put up a building dedicated to ambulatory patients. Charity Hospital on Blackwell's Island also appointed Halsted as a visiting physician, and he was named surgeon-in-chief at the Emigrant Hospital on Wards Island. At the same time, Halsted worked as a visiting physician at Bellevue Hospital. Concerned that conditions at Bellevue did not permit him to operate using his aseptic techniques, Halsted persuaded the commissioner of charities and corrections of New York City to build an operating room for his exclusive use. The facility turned out to be a magnificent tent-like structure on the hospital's grounds, which cost more than $10,000 to erect. He was also appointed to the visiting staff at New York's Presbyterian Hospital and the Chambers Street Hospital. With so many institutions clamoring for his services, Halsted often worked late into the night.

In his six years as a surgeon in New York, Halsted earned a reputation as an audacious surgeon whose operations often pushed the limits of existing knowledge and technique. He pioneered the first autotransfusion to treat laborers who had suffered asphyxiation from working with the gas used to illuminate streetlamps. Halsted treated these patients by bleeding them, shaking their blood in open containers in an attempt to eliminate the carbon monoxide, and then infusing it back into an artery.[13]

Most surgeons are often, and understandably, reluctant to operate on their own family members, but Halsted demonstrated his daring and self-confidence on two memorable occasions. In 1881, his sister Minnie developed uncontrolled bleeding following the delivery of her first baby. Halsted was summoned and managed to stop the hemorrhage, but she had lost a significant amount of blood. Reacting swiftly, he plunged a syringe into his own arm, withdrew blood, and injected it directly into one of his sister's veins—a dangerous move of last resort. She recovered, thanks in large measure to what was probably the first blood transfusion in the United States.

The next year, Halsted was faced with an equally grave family medical crisis. His mother, Mary Louisa Halsted, had become desperately ill. By the time he arrived at her home in the middle of the night, he found her jaundiced, with tenderness in her upper right abdomen. Halsted correctly diagnosed an infected gallbladder. Working by lamplight at 2:00 A.M., he operated on her on her kitchen table, opened her gallbladder, drained the pus, and removed seven gallstones. Once again, Halsted's bold emergency procedure was successful, and his mother regained her health. In a letter to his friend Welch, Halsted later maintained, probably correctly, that this was one of the first surgeries for gallstones performed in the United States.

Halsted's daring and original surgical skills were matched by his gifts as a teacher. Because the regular courses at the College of Physicians and Surgeons didn't prepare students adequately for final examinations, Halsted, like other professors, offered his own private "quiz," often involving ward rounds as well as lectures and demonstrations, as a supplement. His quiz was among New York's most popular, with one student describing Halsted as "an ideal teacher and absolutely at home in his familiarity with his subject."[14]

Even with his exhausting surgical schedule, Halsted found time for friendships and a busy social life. He lived on 25th Street between Madison Avenue and Fourth Avenue (now Park Avenue South) with his friend Dr. Thomas McBride, described by contemporaries as handsome, rich, and free spending. The pair entertained regularly, hosting gatherings of other successful young professionals at dinners and concerts, and even sponsoring serious intellectual discussions on a broad range of contemporary topics. Halsted's good friend and future mentor William Welch was a constant presence at these gatherings. During these halcyon New York years, Halsted was described by one of his well-heeled friends as "a delightful companion."[15]

Addiction and the Struggle for Rehabilitation

While asepsis and the ability to minimize bleeding were two of the foundations of modern surgical technique, the third pivotal innovation was effectively controlling pain. Prior to the introduction of anesthesia, surgery was a rough, crude, and hurried business. With the patient usually awake and at best dulled with whiskey to the piercing agony produced by surgical blades and bone saws, speed was of the essence. As described by the surgeon Emil Goetsch, "Cyclonic surgery was the order of the day—surgeons watching clocks and proudly announcing 'six minutes' or 'ten minutes' when work was completed."[16] The Scottish-born surgeon Robert Liston (1794–1847) was reputed to be one of the fastest surgeons of his day. In one infamous operation, he amputated a leg in less than two and a half minutes. In the process, however, Liston accidentally cut off the fingers of his assistant. Both the patient and the assistant died from infections.[17] In at least one account of this surgery, a spectator was nicked by Liston's knife and subsequently died of shock. Such outcomes, resulting largely from hasty, slipshod, and unsanitary operations, were completely avoidable in Halsted's view.

On October 11, 1884, Halsted read an account in the *Medical Record* by Dr. Henry Noyes, who reported on striking news from the Ophthalmological Congress in Heidelberg, which Noyes had recently attended.[18] Conference attendees had witnessed a convincing demonstration that a 2 percent solution of cocaine, when applied topically to the eye, had a powerful anesthetic effect. Halsted immediately grasped the surgical implications of this finding. Together with his colleagues in New York, he embarked on a series of innovative experiments that demonstrated the ability of injected cocaine to block the sensation of pain in any peripheral nerve and led to the development of local and regional anesthesia. He used himself, his colleagues, and medical students as guinea pigs,[19] clearly ignoring the last passage in Noyes's report noting that "it remains, however, to investigate all the characteristics of this substance, and we may yet find that there is a shadow side as well as a brilliant side in the discovery."[20] Sadly, Halsted and several of his volunteer subjects succumbed to addiction. A number of them died. The eternal search for painless surgery had finally reached its end, but at an enormous and lasting personal cost to Halsted and the others. As the medical historian Howard Markel describes it, "In a matter of weeks Halsted and his immediate circle were transformed from an elite cadre of doctors into active cocaine abusers. Tragically, some of the medical students, resident physicians, and surgeons who participated in these experiments were decimated by the drug and died early deaths."[21] Halsted wrote only once, in the *New York Medical Journal* in 1885, about local and regional anesthesia.[22] His first sentence, rambling incoherently for more than one hundred words, marks it as the product of Halsted's disorienting addiction.

Shaking his cocaine addiction proved difficult, and the available evidence suggests that Halsted was ultimately unsuccessful. In February 1886, Halsted, perhaps accompanied by Welch, boarded a sailboat bound for the Windward Islands in the Caribbean. He carried with him only half the amount of cocaine he would need to maintain himself on the voyage, hoping to cure his habit before returning to New York. Some reports indicate that, unable to bear being without the drug, Halsted broke into captain's medical stores to get hold of a narcotic, probably morphine, for the trip home. With his professional life in New York in shambles and at the urging of his father, his brother, and Welch, Halsted checked himself into the Butler Hospital in Providence, Rhode Island, a facility for the mentally ill and "inebriates."[23] His caregivers deliberately avoided the New York sanatoriums, where his family name was well known, registering the new patient as "William Stewart."

At Butler, in an effort to douse the fires of his addiction, Halsted was apparently treated with morphine. In those days, morphine was not an illegal drug and could be purchased at pharmacies for many common ailments. Sadly, these morphine treatments only led to Halsted becoming addicted to a second narcotic.[24] In November 1886, after a seven-month stay at Butler, Halsted was discharged and accepted an offer of a position from his friend William Welch, who by then had been named chief of pathology at the nascent Johns Hopkins School of Medicine.

Halsted worked as a graduate student in Welch's newly constructed laboratory along with other bright young researchers, most notably the anatomist Franklin Mall. In collaboration with Mall, Halsted applied the European scientific model of medical investigation he had absorbed during his studies there to solve a major surgical challenge. The problem was how to join surgically transected sections of the intestine. Surgeons could remove diseased sections of bowel, but all too often, when they attempted to suture the remaining sections of bowel together, the sutures gave way, spilling intestinal contents into the abdominal cavity. In those cases, the patient would frequently die a horrible, painful death from infection. Halsted, working together with Mall, demonstrated that the fibers that make up the inner lining of the intestine, the submucosal layer, give it unique strength and resiliency.

In a series of ingenious experiments relying on microscopic inspection of intestinal tissues, they demonstrated that sutures would hold if they were placed in the submucosal layer of the intestine. Because of this advance, intestinal surgeries could be performed with much less risk of rupture. Rapid strides in the late nineteenth and early twentieth centuries in alimentary tract surgery were a direct result of Halsted's scientific investigation of this important surgical problem. The work was significant enough that Halsted was invited to present his research findings at Harvard Medical School. Despite the breakthrough and only one month later, in March 1887, he was readmitted to Butler Hospital, still not cured of his addiction.

The Baltimore Years: Extraordinary Surgical Success

Halsted returned to Hopkins nine months later in December 1887—ostensibly cured of his addiction. He resumed his laboratory work, but this time he was also permitted to see patients and perform operations. When the Johns Hopkins Hospital opened in 1889, Halsted was initially appointed to the lesser post of acting surgeon. Still, his remarkable contributions to surgery could not be ignored. In 1890, Halsted was appointed surgeon-in-chief of the hospital and professor of surgery in the medical school. That his addiction was thought to be cured is apparent in the letter William Osler, chief of medicine, wrote to university president Daniel Gilman recommending Halsted for the positions. Osler informed Gilman that "Halsted is doing remarkable work in surgery, and I feel his appointment to the University and Hospital would be quite safe."[25]

Halsted remained chief of surgery and professor of medicine for more than three decades until his death in 1922. A later chief surgeon at Hopkins, the expert biographer John Cameron, writes, "The monumental productivity of Halsted during this period likely never again will be duplicated in American surgery."[26] Halsted revolutionized his specialty by painstakingly applying scientific research to entirely new fields of surgery.

Halsted's conception of himself as a scientist is integral to his indelible legacy. For all of his bold firsts in the operating room, Halsted understood that painstaking research and the steady accumulation of deep anatomical and physiological knowledge were the only genuine guarantee of permanent advances in surgical treatment. Nowhere was this more evident than in his early research with the anatomist Mall to discover the best technique for surgically removing and suturing sections of the intestine. Many of Halsted's other surgical breakthroughs, ranging from delicate operations on the thyroid gland to technical innovations such as the use of silver foil, with its bacteria-killing properties, to cover wounds, were built on extensive laboratory research. Even his discovery of cocaine as an effective nerve-blocking anesthetic was a case of carefully planned experimentation (using himself and his colleagues as guinea pigs) gone terribly awry. Ironically, at the end of his life, Halsted's research on the nerve-blocking properties of cocaine won him belated recognition from the dental profession for clearing the path to painless dentistry.

Harvey Cushing, Halsted's student, who would go on to become the father of neurosurgery, observed in Halsted's obituary, "He had that rare form of imagination which sees problems, and the technical ability combined with persistence which enabled him to attack them with promise and a successful issue."[27] We risk overlooking one of Halsted's lasting contributions to medicine if we fail to take into account his consistent dedication and remarkable record of contributions to surgical research. We have noted that William Welch had largely abandoned his own laboratory research as his very public and multifaceted career as an advocate for scientific medicine thrust him into diverse roles at Hopkins and beyond. Surprisingly, William Hallsted's career took a different path. Known early on as a bold, risk-taking surgeon, Halsted's lasting fame and reputation are closely connected with the groundbreaking research he did to open new pathways in surgery.

If meticulous scientific research was one pole of Halsted's legacy, the other was carefully planned and deliberately executed surgery. Halsted focused on safe surgery, practicing and advocating for it tirelessly at a time when many surgeons continued to operate quickly (even though ether and chloroform had been available as general anesthetics for decades) with little concern for blood loss and careful tissue handling. By the late nineteenth century, rough, hasty work with the scalpel was more a matter of tradition than necessity. Senior surgeons sat atop the pyramid in the medical profession, and their trainees often were reluctant to break the mold of generally accepted operating room practice.[28] In marked contrast, Halsted was meticulous in his surgeries. He encouraged the gentle handling of tissues and preached that "good surgery . . . was bloodless surgery."[29] He routinely tied off "even the most minute vessels—perhaps several hundred in a single operation."[30] If Halsted's surgeries were bloodless, they could also be slow, often painfully so. After observing Halsted operate, Charles Mayo, one of the founders of the Mayo Clinic, is said to have quipped, "It was the first time I ever saw the upper end of an incision heal before the lower end was closed."[31] Because he constantly applied science to important surgical problems and practiced slow, safe surgery, Halsted compiled a "staggering" list of pioneering achievements, according to one of his biographers.[32]

Both as a scientific researcher and as a practitioner of safe surgery, Halsted had a relentless work ethic. In his obituary, Cushing commented that Halsted "spent his medical life avoiding patients . . . when this was possible . . . and, when health permitted, working in clinic and laboratory at the solution of a succession of problems which aroused his interest."[33] (Fittingly, Halsted's last publication, written only months before his death in 1922, returned to the topic of intestinal sutures, one of the first research problems he worked on when he arrived in Baltimore.)

Halsted's struggle with addiction made him reclusive at times, but his choice of research and clinical work over the more collegial aspects of medicine favored by many of his peers was as much a matter of choice as it was of concealment. Emil Goetsch, like Cushing, a student and later a colleague of Halsted's, recalled a conversation with Halsted about surgical residents who chose to marry at a young age. Halsted reportedly said, "'Can you imagine a situation such as this: It is five o'clock and you are in the laboratory at work on a problem that will occupy you far into the night. The telephone rings. It is your wife saying you are expected for dinner as guests are coming. Goetsch, can you imagine?' He [then] shook his head sadly."[34] The story may explain why Halsted himself waited until he was thirty-nine to marry.

Returning to Halsted's many surgical breakthroughs, we have seen that he developed the suturing technique for reliably rejoining two sections of the intestine that opened a path to innovative alimentary tract surgery. In addition to groundbreaking surgical treatments for tumors of the intestine, Halsted used his careful scientific study of anatomy and pathology to advance other untested frontiers. He was the first surgeon to successfully perform bile duct surgery, to remove a periampullary cancer (a cancer that arises at the junction where the bile and pancreatic ducts join the small intestine), and also the first to resect an aneurysm of the subclavian artery. He has been credited with the earliest use of a metal plate and screws to repair long bone fractures and with being one of the first to treat an inguinal hernia (where a loop of intestine protrudes through the abdominal muscles) surgically. Halsted even had the deep anatomical knowledge and meticulous surgical skills to operate on the thyroid gland, a procedure long thought to be too dangerous because of the many nerves and blood vessels that are adjacent to the thyroid. Self-effacing in the extreme during his Baltimore years, he was nonetheless responsible for a vast expansion of modern surgical horizons.

Nowhere was Halsted's skill and research-based approach more apparent than in the surgical treatment of breast cancer. In the late nineteenth century, women with breast cancer faced especially grim prospects. As their tumors grew, they could only watch helplessly. Many tumors grew so large that the cancer ulcerated through the skin. Some patients chose to withdraw from society, while others were faced with horrific breast amputations or hideous cauterizations.

Halsted took a different tack. Using the scientific method, he studied the biology of cancer; he wanted to understand the cells and tissues that contributed to cancer's spread and apply this knowledge in his surgery. He developed an operation that relied on removing the breast, the surrounding lymph nodes, and several chest muscles without cutting into the tumor and risking the spread of the cancer cells into the patient's wound.[35] The results of this procedure, though they were terribly disfiguring, finally gave women hope of surviving a devastating disease. Halsted's procedure was in fact radical: To avoid the possibility of spreading the cancer, Halsted cut widely and only into healthy tissue. As the historian Ric Cottom has said of Halsted's approach, "Yes, it was radical. And, yes, it disfigured. But, most of the time, it worked."[36] With the older, less precisely mapped surgery for breast cancer, recurrence at the surgical site occurred about 60 percent of the time. Once Halsted introduced his radical mastectomy, the local recurrence rate dropped to 6 percent.[37]

While it may have been appropriate for its time, the "Halsted radical mastectomy" is far too drastic and far too disfiguring for today's approach to treating breast cancer. Small cancers can now be detected with mammography, and radiation therapy can improve local control of the disease. Sadly, many surgeons, reluctant to abandon Halsted's practices even though he was long dead, continued to perform this radical breast surgery until a few decades ago.

Halsted was one of the first surgeons in the United States to use rubber gloves in the operating room. Halsted's chief surgical nurse, Caroline Hampton, came to Hopkins Hospital in 1889 (figure 5.3). With his insistence on a sterile operating room environment, Halsted required his assistants to thoroughly wash their hands with permanganate, hot oxalic acid, and sometimes even mercuric chloride before an operation. Hampton developed contact dermatitis and eventually severe eczema from the harsh chemicals. She told Halsted she planned to resign due to the pain. Not willing to lose an "unusually efficient" assistant, Halsted sent plaster casts of Hampton's hands to the Goodyear Rubber Company and asked them to design rubber gloves for her.[38] The first versions Goodyear produced were thick and bulky, more like gauntlets than gloves, but they protected Hampton's hands. On returning from a vacation, Halsted was surprised to find other members of his surgical team using gloves. Once the use of rubber gloves became widespread, the rate of infection from surgical procedures, both among patients and medical staff, dropped dramatically (figure 5.4).

FIGURE 5.3. *Caroline Hampton, who later married Halsted, circa 1889. Halsted developed surgical gloves to protect her from the harsh chemicals surgeons and nurses used to disinfect their hands before operating.*

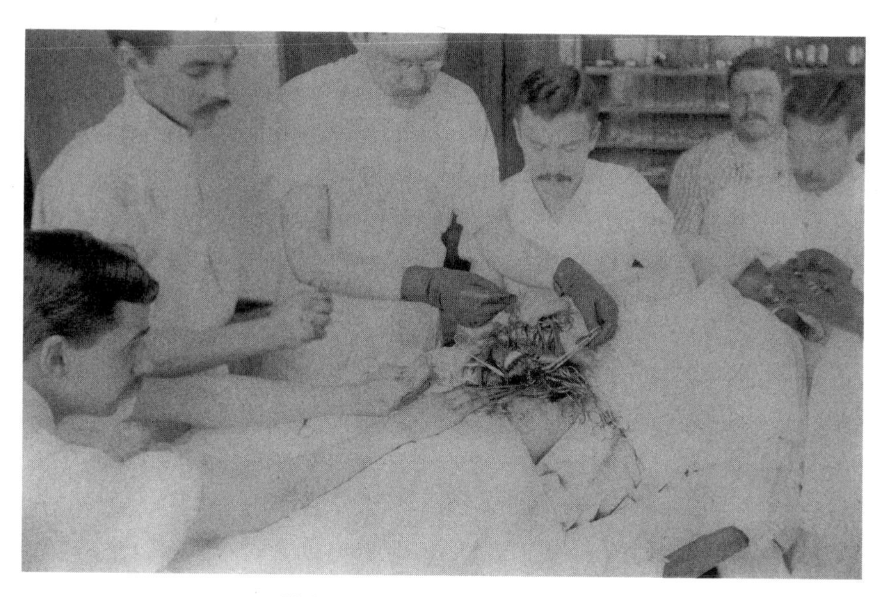

FIGURE 5.4. *Halsted operated slowly, with careful attention to controlling bleeding and gentle handling of tissues. This photo of an 1893 operation is one of the earliest in which Halsted wore gloves.*

There may have been an element of romance involved in developing gloves to protect his surgical nurse, because Halsted married Caroline Hampton in 1890. The daughter of a prominent South Carolina family, Hampton resisted a traditional role in southern society. She struck out for a career of her own in nursing. Heading north to New York City to seek training, she graduated from the nursing training program at Mount Sinai in 1888, and in 1889 was hired as chief of surgical nursing at Johns Hopkins.

Perhaps Halsted's crowning achievement was adopting a formal system to train the next generation of surgeons. The system was similar to the one William Osler introduced at Hopkins to train residents in the Department of Medicine.[39] Before Johns Hopkins Hospital opened in 1889, surgeons in the United States were largely self-taught or at best spent a year or two apprenticing in a hospital.[40] At Hopkins, Halsted set up a new regime in which medical school graduates were educated in a hospital-based, university-sponsored surgical training program that typically lasted six or seven years. Over the course of the program, these future academic surgeons were expected to acquire a solid grounding in anatomy, pathology, and physiology and to take on increasing responsibilities. In the final stage of the program, the seasoned trainees, usually as "resident surgeons," earned full operating-room independence. Halsted's surgical residency training model and its Hopkins graduates gradually spread across the country and eventually populated a number of leading hospitals with their future surgical chiefs. "It is this method for training surgeons introduced by Halsted," according to Cameron, "that probably is more responsible than any other single factor for the incredible productivity that has placed the United States in the forefront of surgical science throughout the world."[41]

The importance of developing a model for training surgeons who practiced safe surgery and were deeply versed in the basic disciplines central to human biology cannot be overestimated. But in Halsted's time, the most immediate and consequential effect was to seed the profession with a cadre of scientifically minded disciples. In addition to supplying hospitals around the country with their surgical chiefs, Halsted equipped a good number of his trainees, including Harvey Cushing in neurosurgery, Hugh Young in urology, and Samuel Crowe in head and neck surgery, with the skill and drive to establish new surgical subspecialties. It was no accident that in the early twentieth century institutions like Harvard, Yale, Cornell, Stanford, and Pittsburgh all had professors of surgery who had been Halsted's chief residents. These men (predictably for the times, there were no women) set up surgical training programs modeled on Halsted's vision.

There is some irony in praising Halsted as an educator. In New York, he was renowned for his innovative private "quizzes." But in Baltimore, even his strongest supporters admitted that Halsted wasn't a good teacher. One colleague called Halsted's bedside discussions of patient cases with his students "a rather laborious proceeding."[42] In his obituary in *Science*, Harvey Cushing noted, "A bed-to-bed ward visit was almost an impossibility for him. If he was interested he would spend an interminable time over a single patient"[43] (figure 5.5). Rounds with Halsted were even described as "shifting dullness."[44] In fairness, things were quite different for Halsted's residents, who encountered him as a mentor and a trainer of surgeons rather than as a bedside teacher. As William MacCallum pointed out in his biography of Halsted, in those more direct and personalized circumstances, where "it is not a question of instruction but rather of inspiration . . . there are a few such men [as Halsted] in the world."[45] Halsted may have been an ineffective teacher in Baltimore, yet he ended up developing and inspiring an extraordinary generation of surgeons (figure 5.6).

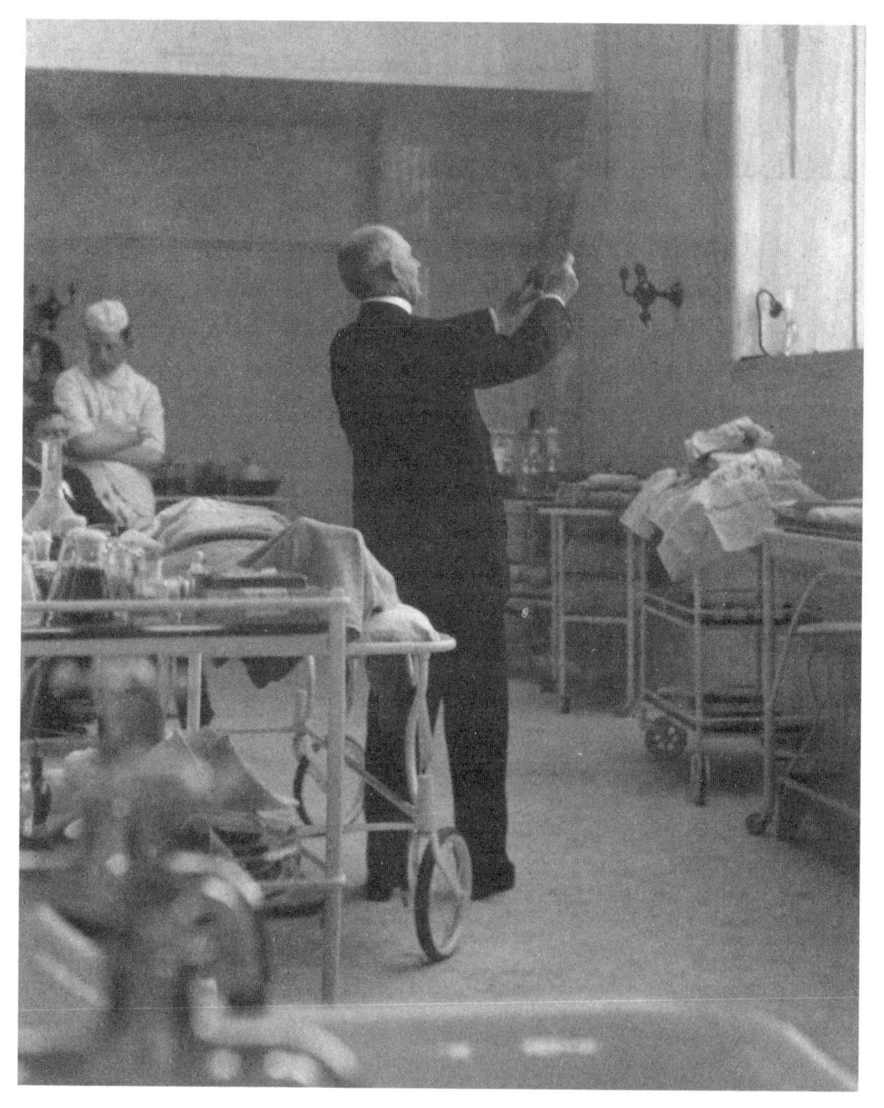

FIGURE 5.5. *Halsted examining an X-ray on teaching rounds, circa 1904. In contrast to Osler's rounds, Halsted's were described as slow and interminable.*

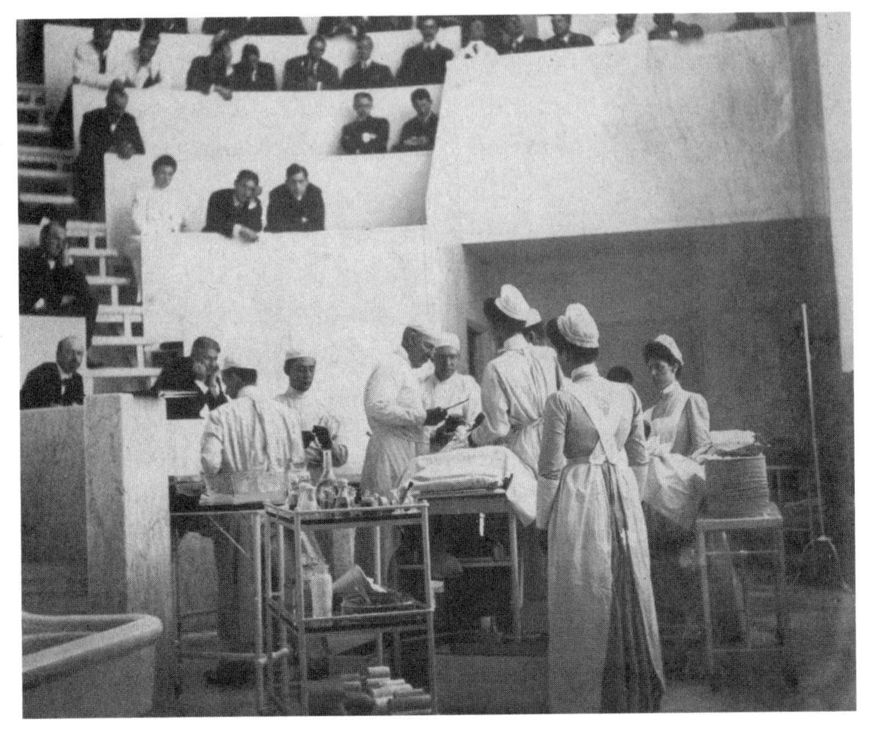

FIGURE 5.6. *Halsted, standing at the operating table with hands gloved, taught by example. It is hard to imagine what the observers in the top rows of seats could actually see. (Photo is from 1904.)*

The Price of Secrecy

We have alluded to the marked contrast between the bold, daring, and gregarious Halsted during his years in New York and the reserved, fastidious, and secretive figure friends and colleagues observed after he arrived in Baltimore. Halsted's reclusiveness was likely the result of his inability to cure himself of his morphine addiction. In a secret diary only opened decades after his death, William Osler wrote, "About six months after the full position had been given, I saw [Halsted] in a severe chill, and this was the first intimation I had that he was still taking morphia. Subsequently I had many talks about it and gained his full confidence. He had never been able to reduce the amount to less than 3 grains daily, in this he could do his work comfortably and maintain his excellent physical vigor. . . . I do not think anyone suspected him—not even Welch."[46] In his compelling and well-written book *An Anatomy of Addiction*, the medical historian Howard Markel makes a convincing case that even Halsted's decision to organize his illustrious residency training program was fueled by the surgeon's ongoing dependence on narcotics. After all, Markel argues, what better way to cover up and explain his apparently frequent drug-induced absences from the operating room and the wards than by "sitting atop a pyramid of eager young doctors willing to stay up to all hours tending to his patients."[47] Whatever its ramifications on his life and his work, Halsted's continued addiction remained a secret from all but a very few who knew or thought they knew him.

Halsted led a quiet, mostly private life in Baltimore. In his first years there, he frequented the Maryland Club, often dining with Welch, who was one of the only people he allowed to become close. After marrying Caroline Hampton, the couple maintained an elegant home at 1201 Eutaw Place in Baltimore's fashionable Bolton Hill neighborhood. They seem to have used its ample entertainment rooms more for formal occasions than casual socializing. One longtime acquaintance noted that he had visited the Halsted's home only a couple of times. Hampton was the niece of a famous Confederate general, Wade Hampton. In addition to their Baltimore home, the couple purchased a rural retreat they dubbed High Hampton in the mountains near Cashiers, North Carolina. The property had been the Hampton family's hunting lodge, and the newly married Halsteds had honeymooned there. Caroline was an accomplished horsewoman. Halsted spent his time gardening and pursuing his interest in astronomy. He was one of the country's leading growers of dahlias.[48] The Halsteds had no children but lavished their attention on their dogs, two of which were perfectly named: Nip and Tuck (figure 5.7).

Perhaps to shield himself from unwanted scrutiny, Halsted was fastidious in his outward appearance and social routines. He was an elegant dresser, frequently sporting a silk top hat or a derby. His suits were tailored in London, his shoes were made in Paris, and his shirts also sewn there.[49] Insisting that he was unable to find an adequate laundry service in Baltimore, Halsted was in the habit of sending his shirts back to Paris to be laundered. He disliked having his photo taken, as evidenced by the relatively few photos we have of him outside the operating room or with medical colleagues.

FIGURE 5.7. *A rare informal photo of Halsted at his beloved country home in Cashiers, North Carolina, where he spent part of his summer each year with his wife Caroline. His dachshunds were appropriately named Nip and Tuck.*

When the couple entertained, mostly early in their marriage, Halsted orchestrated their dinner parties with the same meticulous care he applied to his surgery. He personally supervised buying the food for his guests, selected the individual coffee beans to be used, and chose the logs for the fireplace. Halsted's specifications were for "hickory aged three years in a dry place . . . and shipped from . . . North Carolina."[50] Even the tablecloths didn't escape Halsted's attention. He made certain that the linens were ironed after they were spread on the table to prevent any creases. As the years went on, the Halsteds withdrew from socializing in Baltimore. Caroline Halsted had retired as a surgical nurse after her marriage, as nurses in that era were expected to do. She began to spend several months a year starting in the spring at their beloved High Hampton country home, where Halsted would join her each summer after returning from visits with medical colleagues in Europe.

One thing that didn't change was Halsted's continuous record of surgical accomplishment. As late as 1920 and entering his fourth decade as chief of surgery at Hopkins, Halsted published a 186-page, single-author monograph on surgery for goiter.[51] In his authoritative and engaging article about Halsted, John Cameron notes that Halsted received honorary degrees from Yale, his alma mater, and Columbia. Halsted also had the singular distinction of being inducted into the National Academy of Sciences. For the most part, however, his monumental contributions to surgery were underrecognized by his peers. Halsted's reclusive demeanor and habit of burying himself in his work undoubtedly contributed to this lack of public recognition.

At a deeper level, we inevitably come back to the clear differences between the bold, risk-taking, and socially gregarious Halsted in New York and the reserved, meticulous, and intensely private figure of the Baltimore years. In the sealed diary entry mentioned earlier, Osler wrote, "The proneness to seclusion, the slight peculiarities, amounting to eccentricities at times, (which to his friends in New York seemed more strange than to us), were the only outward traces of the daily battle through which this brave fellow lived for years."[52] It is easy to take Osler's description of Halsted's struggles at face value. His morphine addiction, which by Osler's account continued at least through 1912, was certainly a huge burden to shoulder. Yet, as Cameron argues, it is also likely "that Halsted's immense academic and scholarly productivity was not despite his addiction, but in part because of it."[53] After his second effort at rehabilitation in Butler Hospital, Halsted moved to Baltimore and rebuilt his life. Working in Welch's pathology laboratory and then gradually resuming his surgical career enabled Halsted to fashion a new life that allowed him to conceal his secret from all but his closest friends and the time and space to focus intensively on pathbreaking surgery and research. "Had he not developed his disability," Cameron concludes, "it is unlikely he would have followed Welch to Baltimore and Hopkins. The chance to work with brilliant colleagues in a new hospital associated with the first research and graduate university . . . in this country would have been lost . . . An unusual set of circumstances led Halsted to . . . the Johns Hopkins Hospital, and the result was an amazing productivity that led to the creation of our surgical heritage."[54] In a similar vein, the neurosurgeon Harvey Cushing, who worked closely with Halsted, suggested that "it might even seem that the whole Halsted school of surgery which I have called a School for Safety in surgery may have been due to this drug addiction." Cushing added that as a result of his addiction Halsted's "outlook on surgery itself was changed and he in turn devoted himself to the infinite precision of little details of surgery."[55]

Like most secrets, Halsted's addiction exacted a terrible personal price. That price was multiplied many times over, as Markel notes, by the risk that detection posed to Halsted's "illustrious reputation." (Like Welch's probable homosexuality but perhaps even more so because it directly impaired his ability to complete his surgical duties, Halsted's addiction required an immense effort to conceal.) In the final chapter of his book, Markel describes the stark, almost inconceivable, split between the surgeon who by day would "focus intensely, if not obsessively, on advancing the craft that would—and continues to—save millions of lives" and the addict who "when the clock struck four-thirty . . . hurried home . . . took out . . . a soothing dose of morphine . . . and sailed off to narcotized oblivion."[56] The relief could only have been momentary and bittersweet at best, given Halsted's fierce devotion to his work.

Throughout the four decades of his drug dependency, Halsted remained, in Markel's view, "a remarkably high-performing addict," largely able to confine his drug binges to periods outside of hospital routines and to alternate these secretive episodes with intense bursts of surgical productivity.[57] Equally remarkable was the fact that Halsted was able to hide this antipodal divide in his life "for so long, while so few knew of this habit."[58] The tragedy, as Markel eloquently puts it, is that "the ashamed, guarded, and lonely Halsted concealed this part of his life to the very end."[59] Modern psychology recognizes the immense cost of secrets to the self and loved ones. William Halsted lived and worked in an era in which hiding a secret as shameful as addiction, whatever pain and loneliness it entailed, must have seemed his one and only choice.

Death and Remembrance

Halsted's life came full circle at the end. Decades earlier, as a fearless young surgeon, he had operated on his desperately ill mother in the middle of the night to remove gallstones and also had transfused his own blood into his hemorrhaging sister to save her life. In 1919, one of Halsted's former residents, Richard Follis, operated on Halsted to remove his gallbladder and clear stones from his bile duct. Three years later, in the summer of 1922, Halsted was at his summer home in North Carolina and again began to experience symptoms of bile duct obstruction. He traveled back to Baltimore where George Heuer, another one of his former trainees, operated on him on August 25 to remove a gallstone lodged in his bile duct. Nine days later, Halsted began to bleed internally. One by one his surgical residents lay down beside him to transfuse their blood. Dr. William Rienhoff Jr. was among those residents who gave his blood to Halsted. One of Rienhoff's sons later recalled, "This was obviously a difficult thing for all these people who were so close to him to have to do. They all had a great affection for him and a great respect for him."[60]

The effort was to no avail. Halsted died thirteen days after his surgery on September 7, 1922, two weeks shy of his seventieth birthday. His ashes were interred in Greenwood Cemetery in Brooklyn after a small private funeral at the Halsted's Eutaw Place home in Baltimore, presided over by his roommate from Yale, the Reverend Samuel Bushnell. Only Halsted's brother and two sisters and their families, along with Halsted's friend of more than four decades William Welch, attended the burial. Caroline Halsted was too ill and grief-stricken to make the journey from Baltimore. Fittingly, perhaps, it was a private and quiet end to Halsted's brilliant but tortured career.

Halsted's obituary in the *Baltimore Sun* assessed his legacy this way:

> Because Dr. William S. Halsted lived the world is a better, a safer, a happier place in which to be. In his death not only Baltimore, but civilization as a whole, has sustained a heavy loss. He was one of the few men who really count. Quiet, simple, unostentatious except in the medical world, where he towered a great and dominating figure, the full scope of his genius and the tremendous extent and value of his services to mankind were neither generally known nor generally appreciated.[61]

Of necessity, we must look back at William Stewart Halsted's life from our twenty-first-century vantage point, a vantage point shaped both by far more advanced, technologically informed surgical methods and by a deeper and more compassionate understanding of addiction (color plate 9). That said, we cannot help but be struck by two facets of Halsted's legacy. The first is his unwavering, constantly renewed dedication to applying research and scientific methods to the great surgical problems of his day. As a result of his work, surgery became a science, and the epicenter of surgical innovation moved from Europe to America. The second is Halsted's undaunted courage and relentless will to move forward with his work in the face of the crushing burden of morphine dependence. Howard Markel summed it up best when he wrote of Halsted and Sigmund Freud (who also suffered from cocaine addiction), "in defiance of the malady that nearly destroyed them—or perhaps because of their struggle to overcome it—neither man ever lost his zeal for delivering his healing gifts to the world."[62] Halsted's story is like that of the mythical phoenix rising from the ashes of its own death. Halsted emerged a different person from his addiction, but one who forever changed surgery (figure 5.8).

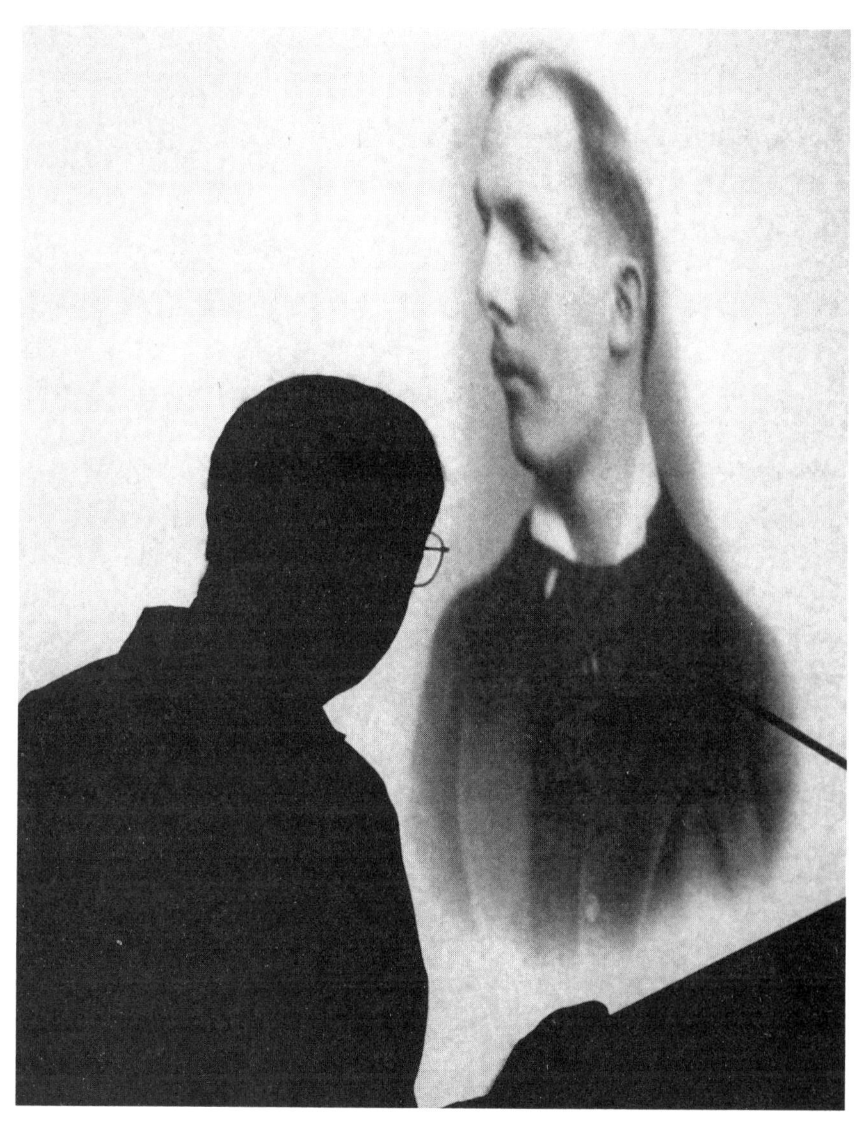

FIGURE 5.8. *John Cameron (foreground), the fifth chair of the Department of Surgery at Johns Hopkins, presenting a talk on Halsted. Halsted's inspiration continues to be passed down to trainees today. (Photograph by Jed Kirschbaum.)*

6

JESSE WILLIAM LAZEAR

The Ultimate Sacrifice

The fever chart recorded a temperature of 102 degrees on September 19, 1900, the day the patient was admitted to the yellow fever ward. The chart carried the name of Dr. Jesse W. Lazear, the thirty-four-year-old Hopkins physician lying gravely ill in the US Army's Columbia Barracks hospital near Havana (figure 6.1). By the next morning, Lazear's temperature had climbed to 104 degrees, dangerously high for a young man who only days earlier had gone for one of his frequent evening swims in the nearby ocean.

His fever broke suddenly on the night of September 23, but his medical condition continued to worsen. Like most patients in the advanced stages of yellow fever, Lazear's illness progressed quickly. He soon experienced convulsions, delirium so severe he had to be restrained, and the dreaded *vomito negro*. "Black vomit" was in fact the name terrified Spanish colonizers of the Yucatán had given the disease when they first encountered it in 1648. Two days later, at 8:45 P.M. on September 25, Lazear was dead.

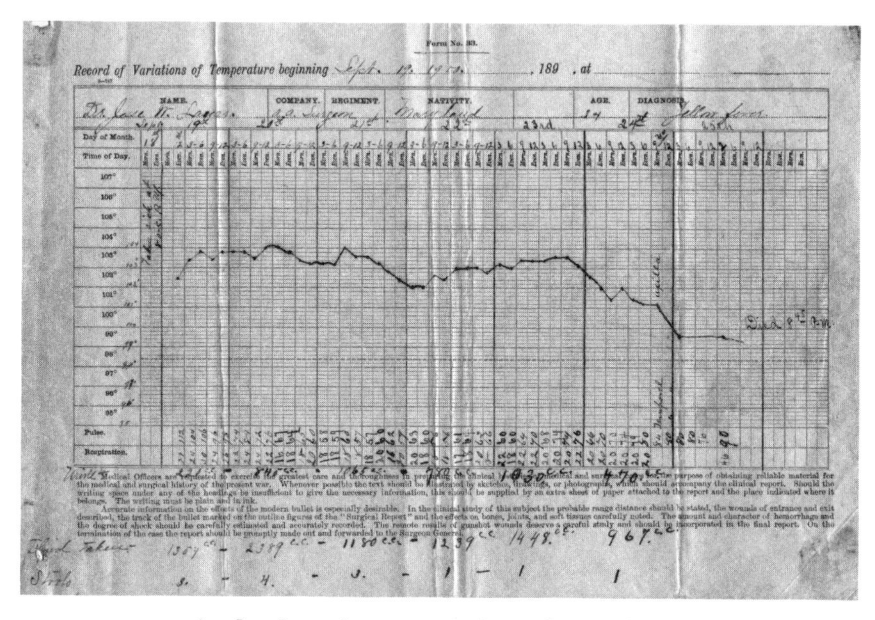

FIGURE 6.1. *Jesse Lazear's temperature chart as he struggled with yellow fever. The chart ends with the curt notation "Died 8:45 P.M."*

Lazear's death was tragic. He left behind a young wife and two young children, one of whom had been born only two weeks earlier. His passing cut short a promising career. Lazear had been head of the clinical laboratory at Johns Hopkins Hospital and, as a trainee of William Welch, had been handpicked to join the US Army's Yellow Fever Board, commissioned in the wake of the Spanish-American War to identify the cause of the disease and eradicate it once and for all from the newly won territory of Cuba. In another sense, however, Lazear's death reflected the determination of a young man to risk everything in the pursuit of knowledge. It also provides a window into the relentless, if sometimes slow and circuitous, progress of scientific medicine. Finally, Lazear's premature death brings into sharp focus the devastation, fear, and racial discrimination that yellow fever, an unseen and unknown killer, brought with it before he and his colleagues put the Americas on a path to controlling it in the first years of the twentieth century.

A Horrible but Elusive Killer

Prior to Lazear's death, the cause of yellow fever and the mechanisms by which it spread were unknown. The principal suspected modes of transmission—supported more by superstition and prejudice than by hard empirical evidence—were miasma (bad air) or "fomites." The latter are objects and materials that had come in contact with the blood, feces, urine, or other effluvia of yellow fever victims. With so little known, yellow fever had long sowed terror and bewilderment among the defenseless populations who encountered its ravages.

Today, thanks in large part to Lazear's daring work, we know that the disease is spread by the bite of the female *Aedes aegypti* mosquito. Microbiology was still in its infancy early in the twentieth century when Lazear and his Yellow Fever Board colleagues identified the disease-carrying insect, or vector. We now understand that the pathogen transmitted from an infected human by the mosquito's bite is a flavivirus, closely related to the viruses that cause West Nile disease and dengue fever. The flavivirus that produces yellow fever is an organism so tiny as to escape routine microscopic detection.

When Jesse Lazear joined the battle against yellow fever by volunteering to serve on the army's Yellow Fever Board, the disease, which originated in Africa, had been a scourge in the Western Hemisphere for over 250 years. As early as 1648, Diego López de Cogolludo, a Spanish missionary to present-day Mexico's Yucatán region, described the first recorded outbreak of yellow fever in the Americas. Those afflicted, according to the Spaniard's graphic account, were "taken with a very severe and intense pain in the head and of all the bones of the body, so violent that it appeared to dislocate them or to squeeze them as a press . . . [and a] vehement fever" frequently combined with delirium. The truly unfortunate began "vomiting putrefied blood [*el vomito negro*] and of these very few remained alive."[1] Cogolludo's narrative was the first but definitely not the last we have of yellow fever's devastating impact in the New World.

The Slave Trade's Evil Twin

The spread of epidemics often follows major routes of commerce or migration, and yellow fever's journey to the Americas was no exception. The rapid expansion of the slave trade in the late seventeenth century to meet the demands of the Caribbean's plantation-based economy for forced labor opened multiple transatlantic sea lanes for the spread of yellow fever. In addition to their forcibly boarded human cargo, slave ships plying the Atlantic took on stores of sweet potatoes and other native foods to feed their captive passengers before they departed Africa's west coast for the three-month crossing. They also took on casks of water for the long journey. The casks were fertile hatching grounds for mosquitoes. Within days of leaving port, larvae would emerge below the surface, followed in another few days by full-grown mosquitoes. A slave ship's crew and human cargo were a ready source of the blood meals the females depended on to return to the casks and lay their eggs. During a single ocean crossing, generations of mosquitoes could be hatched.

As authors John Pierce and Jim Writer point out in *Yellow Jack*, their excellent history of yellow fever in the Americas, it is impossible to know the precise toll the disease took on enslaved people brought to the West Indies. But considering the estimated 20 percent mortality rate from multiple sources, including disease, murder, and suicide, one thing is almost certain: "only the mosquito population had increased on the crossing."[2] If nothing else, the dramatic increase in the population of enslaved people in Barbados in the mid-1600s and the surge in outbreaks of yellow fever there and throughout the Caribbean in the late 1700s support the notion that the Atlantic slave trade was a major contributor to making the disease endemic in the Americas for more than two centuries. From the Caribbean, yellow fever was exported to North America, following the trade in rum, sugar, and, eventually, enslaved people, as forced labor grew into a major economic engine.

Recently, genetic sequencing of the yellow fever virus has confirmed that the virus likely originated in Africa 1,500 years ago. The sequencing data also "show that the spread of YFV [yellow fever virus] to the Americas corresponds closely with the routes and timing of the slave trade."[3] These findings are crucial in understanding the origins and spread of yellow fever, but also carry a stark warning for today, as travel and commerce have become embedded in the fabric of our lives.

A Tale of Two Cities: Philadelphia and Memphis

An outbreak of yellow fever in one of North America's largest cities in the late eighteenth century illustrates the horrors of the disease and just how little was known about its causes and treatment. Accounts of this yellow fever epidemic also illustrate the role that racism played in the often-panicked response to yellow fever. In the summer of 1793, Philadelphia, then the nation's capital, suffered a devastating outbreak of yellow fever. According to a 1998 *Scientific American* article about the outbreak, an estimated five thousand people, or approximately 10 percent of the city's population, died.[4] The epidemic appeared to have been ignited by French refugees from a slave rebellion in present-day Haiti who landed on Philadelphia's wharfs along the Delaware River. Within weeks, working-class residents of the area were stricken with the typical symptoms of high fever, hemorrhages, yellow skin and eyes, and, in the most severe cases, black vomit.[5]

Though initially confined to people on the geographic and socioeconomic fringes of the city, the outbreak soon spread across Philadelphia, causing widespread panic, sickness, and death. Those Philadelphians who could afford to leave began to flee the city, including many of the city's eighty physicians. The poor, of course, were not able to leave and resorted to protective measures recommended by Philadelphia's prestigious College of Physicians, including such fanciful (at least to our modern way of thinking) antidotes as hanging camphor bags or tarred ropes around their necks, stuffing their pockets with garlic, or shooting off guns in the house—presumably to ward off bad air.[6] As the epidemic spread, Philadelphia's social fabric was torn asunder. Spouses abandoned their dying partners. Orphaned children were left starving and roaming the city's streets. Shops and markets closed, and most government activities ground to a halt.[7]

Worst of all, Philadelphia's fragile public health and welfare institutions were overwhelmed. Pennsylvania Hospital and the city's two almshouses closed their doors to yellow fever victims. In desperation, city authorities commandeered a mansion previously occupied for a time by then vice president John Adams and his wife Abigail and converted it into a makeshift hospital. One Philadelphia resident described conditions that were horrific: "The dying and dead were indiscriminately mingled together. The ordure and other evacuations of the sick were allowed to remain in the most offensive state imaginable."[8]

In a perverse twist on racial stereotypes, Black people were thought to be somehow different in their bodily makeup and therefore immune to yellow fever. Benjamin Rush, Philadelphia's most prominent physician and a signer of the Declaration of Independence, put out a call for Black nurses to care for the sick. He wrote the African American minister Richard Allen, explaining, "It has pleased God to visit this city with a malignant and contagious fever, which infects white people of all ranks, but passes by persons of your color."[9] Allen and his fellow minister Absalom Jones organized members of the Black community to nurse the sick and bury the dead. A subsequent account by Allen and Jones credited Black Philadelphians "for saving the lives of some hundreds of our suffering fellow mortals."[10]

The two ministers and their cadre of volunteers put their lives on the line caring for the sick and the dying—not from person-to-person infection but due to the risk of being bitten by a mosquito that had fed on the blood of one of their patients. They ultimately paid a terrible price. Allen himself contracted the disease and nearly died. While almost one-third of Philadelphia's white population fled the city, Allen and Jones convinced most free Black people to stay behind. In the end, Black mortality per capita was every bit as high as the death rate in the general population. Adding insult to injury, the Black community had to endure a pamphlet circulated toward the end of the epidemic accusing its members of theft, extortion, and other crimes committed against whites in their care. All in all, this "Philadelphia story" was a sorry legacy built on a fallacious belief in inherent racial differences.

The Memphis yellow fever epidemic of 1878 occurred nearly a century after Philadelphia's outbreak but produced similar devastation and panic. In the end, it took the lives of over five thousand Memphians, or about one-tenth of the city's population, and was one of the worst yellow fever episodes in North America.[11] As in Philadelphia, doctors were overwhelmed and powerless to cope. Folk remedies, including ingesting castor oil, quinine, or potash solution, wearing bags of supposedly protective herbs around the neck, and even shooting off guns to dissipate foul air, were again tried to no avail. Early in the epidemic, church bells rang day and night to signal yet another funeral, but the practice was ended out of deference to the feelings of the ill and dying. As for those who had already succumbed to yellow fever, undertakers and gravediggers worked round the clock to give them a Christian burial. A contemporary account suggested that it was common to see carts piled high with corpses. When one nurse stopped a cart driver to inquire about a man's address, the driver pointed over his shoulder and declared, "I've got him here in his coffin."[12] As in Philadelphia, there were huge disparities in mortality rates between those who fled the city and those who remained behind. In Memphis, too, these disparities broke down along racial lines. Of the twenty thousand residents who remained in the city, fourteen thousand were Black and only six thousand were white.[13] Despite the passage of nearly a century since Philadelphia's yellow fever epidemic, science had made no progress against this devastating disease.

Flying Blind: Bewilderment about Yellow Fever's Causes and Spread

Castor oil, tarred ropes worn around the neck, and gunpowder explosions were feeble defenses against an invisible mosquito-borne virus. Yellow fever raged on and terrorized scores of North American cities and towns in the eighteenth and nineteenth centuries. A large part of yellow fever's horror was the almost complete lack of understanding of its causes, the conditions that fostered its spread, and, certainly, how to treat such a devastating illness. It was an age in American medicine when ancient medical conceptions of the causes of disease still held sway. At a time when most illnesses were attributed to imbalances in the four bodily humors—blood, phlegm, yellow bile, and black bile—the etiology of yellow fever was mainly a matter of unsupported speculation. Benjamin Rush was an adherent of the four humors notion of disease. He proposed to treat yellow fever by restoring the body's natural liquid balance with emetics, laxatives, bleeding, and making the patient perspire. Bleeding was at the top of Rush's list. He wrote in one account of his efforts to treat yellow fever patients, "Toward the height and close of the epidemic, I saw no inconvenience from the loss of a pint, and even 20 ounces of blood at a time."[14]

In the absence of any understanding of microbes as disease agents or the role of other species in their transmission, doctors cast their suspicions on impurities or vapors in the air ("miasmas") as the culprit in yellow fever. Rush, for one, claimed to know the source of the contagion in Philadelphia. After treating a woman who lived near Ball's Wharf, he declared that the source of yellow fever was the putrid air emanating from a cargo of West Indian coffee that had spoiled in transit and was rotting and stinking in Philadelphia's blazing summer sun.[15]

A competing theory, only slightly more sophisticated because it assigned a specific method of transmission, argued that yellow fever was caused by so-called fomites. As noted at the outset of this chapter, these were the objects and materials that had come in contact with infected people, such as bedding, clothing, or other personal belongings. Neither the fomite theory nor the miasma theory was supported by any systematic evidence or certainly any experimental proof. That left most people adrift in a sea of fear and doubt. One Memphis resident who lived through the epidemic of 1878 commented on the state of medical knowledge at the time: "We really know nothing about yellow fever; that it is a law unto itself in its tenacity of life as well as in its inception, growth, and progress."[16]

Attempts to limit the spread of yellow fever took one of two paths. So-called contagionists, who generally favored the fomite theory and person-to-person transmission, argued strenuously for quarantine of the sick and the suspected sick. Experience with diseases like smallpox and bubonic plague had shown that isolating people with these diseases and travelers arriving from cities suspected of infection could sometimes stem the spread of illness. Incoming ships were held on offshore islands and required to fly a yellow flag (hence the nickname Yellow Jack given to the disease).[17] Passengers were isolated for forty days or longer, along with those arriving by land. (*Quarantine* is derived from *quadraginta*, the Latin word for forty.) Then, as now, quarantine measures were marked by racial and ethnic prejudice. As they are today, these restrictions were opposed by business and commercial interests, who saw them as a barrier to economic recovery.

The second line of attack on yellow fever was preferred by "anti-contagionists" or "environmentalists." The environmentalists favored investment in public works projects that eliminated the filthy conditions that they believed contributed to miasma or putrid air. Sanitation projects like hauling away trash and building municipal sewage and water treatment systems were the usual response to epidemics. These steps helped curb the spread of some diseases but were largely ineffective against yellow fever, unless they happened to drain pools of water where mosquitoes bred. In short, adherents to both competing schools of thought on the origins and spread of yellow fever could point only to anecdotal evidence to support their views. (Benjamin Rush's insistence that putrefying coffee on the Philadelphia wharf caused the city's yellow fever outbreak is a prime example of this kind of reasoning.) With scientific data and evidence almost completely missing in the debate over yellow fever, both sides were engaged in blind speculation. Finally, in the last decades of the nineteenth century, there were signs that things were about to change.

Early Scientific Stumbles

Scientific discovery often advances in a two steps forward, one step backward fashion, with some hypotheses leading down blind alleys and others ignored because they fall outside the conventional wisdom. This was definitely the case in the search for yellow fever's cause and especially its method of transmission. As the nineteenth century drew to a close, the emerging science of infectious diseases offered a promising new perspective. The germ theory developed by scientists like Robert Koch in Germany and Louis Pasteur in France touched off a veritable gold rush to discover the specific microorganisms that could account for particular diseases. Yellow fever was no exception, and by the 1890s, two rival hypotheses had been proposed.

In 1890, the Hopkins-trained pathologist George Sternberg, who will figure prominently in our story, published a report titled "Etiology and Prevention of Yellow Fever."[18] Sternberg mistakenly asserted that the cause of the disease was a microorganism, dubbed *Bacillus X*, which he claimed to have found in the bodies of yellow fever victims. By 1897, a second researcher, the Italian scientist Giuseppe Sanarelli, working in Brazil and Uruguay, published an article in the *British Medical Journal* proposing his own pathogenic agent.[19] He reported that he had found *Bacillus icteroides* in the bodies of both living and deceased yellow fever patients. Sanarelli further claimed that he had inoculated five unsuspecting patients with the bacillus and soon after observed "typical yellow fever."

In a testament to Sanarelli's unscrupulousness, three of the five patients died. William Osler, one of the era's guardians of medical ethics, noted that Sanarelli had not obtained consent from his patients for these experiments. He wrote, "To deliberately inject a poison of a known degree of virulence into another human being, unless you obtain that man's sanction, is not ridiculous, it is criminal."[20] The idea of informed consent—the notion that patients participating in medical experiments should be fully apprised of the risks—would make rapid strides during the struggle to understand and bring yellow fever under control. Sanarelli's experiments on human subjects, however, remain an example of how science should not be conducted.

Despite glaring deficiencies in his ethics and his methodology, Sanarelli's alleged discovery was hailed by the prestigious Pasteur Institute and reverberated throughout the scientific community. The buzz notwithstanding, Sternberg disputed Sanarelli's findings and asked two of his colleagues, Walter Reed and James Carroll, both of whom would later join the US Army's Yellow Fever Board in Cuba, to prove Sanarelli wrong. By 1899, the pair had published a report in *Medical News* establishing conclusively that while *Bacillus icteroides* was present in some yellow fever patients, it was a secondary infection and not a causative agent.[21] As it turned out, Sternberg's proposed candidate, *Bacillus X*, would also be ruled out as the cause of yellow fever (the yellow fever virus was not isolated until 1927), but science frequently advances in fits and starts. Linking yellow fever to a specific pathogen and not to some vague notion of miasma or fomites was an important step forward.

Enter Carlos Finlay

While the debate over a supposed bacterial cause of yellow fever raged on, a persevering researcher was painstakingly investigating the disease's method of transmission. The son of a Scottish physician who owned a coffee plantation and a mother who was French, Carlos Juan Finlay was born in Cuba in 1833. After obtaining his medical degree at Jefferson Medical College in Philadelphia, Finlay returned to his native island to practice general medicine, specializing in ophthalmology as his father did.[22] Widely read and imbued with far-reaching scientific curiosity, Finlay devoted his days to treating patients, but used his free time to set up his own research laboratory, even using the microscope he had acquired as a medical student to examine countless clinical specimens.[23] Finlay's medical imagination ranged far and wide, but yellow fever, along with other tropical diseases endemic in Cuba for centuries, was certainly a natural target of his interests.

Finlay studied the anatomy, physiology, and biting habits of the *Culex* mosquito and recorded his observations meticulously.[24] (The species was later renamed *Aedes aegypti,* and that's what we will call it except when referring to Finlay's work with this species.) Finlay correlated his observations with information he assiduously collected about the species' habitat and preferred climactic conditions, geographic distribution, and breeding cycle. Noting the marked overlap between his behavioral and habitat observations about mosquitoes on the one hand and the reported incidence of yellow fever on the other, Finlay concluded that the *Culex* mosquito transmitted yellow fever.

After completing this brilliant epidemiological work, Finlay took one more critical step. In June 1881, he tested his hypothesis about mosquito transmission by inoculating a small group of volunteer subjects using *Culex* specimens that had previously bitten patients with known cases of yellow fever. (The initial group contained five experimental subjects and fifteen more volunteers as controls.) Finlay's first subject was Francisco Beronat, a twenty-two-year-old soldier in Spain's occupying army in Cuba. Within nine days, Beronat exhibited classic symptoms of yellow fever, including a high fever, jaundice, and albumin in his urine.[25] We know that Finlay obtained "necessary authorization" from Spanish military for his inoculation experiments, but it is unclear how well informed these "volunteers" actually were.[26]

On August 14, 1881, Finlay presented a paper titled "The Mosquito Hypothetically Considered as the Agent of Transmission of Yellow Fever" to Havana's Royal Academy of Medical, Physical, and Natural Sciences. The paper summarized his preliminary findings, including the one unambiguous case of yellow fever among his five initial experimental subjects. After quickly dismissing "any theory which attributes the origin and propagation of yellow fever to . . . miasmatic . . . conditions, to filth or to the neglect of general hygienic precautions," Finlay asserted: "The meteorological conditions which are most favorable to the development of yellow fever are those which contribute to increase the number of mosquitoes."[27]

Finlay proceeded to add to this groundbreaking conclusion the more qualified evidence from his experimental inoculations, which had resulted in a single confirmed case of yellow fever. A scrupulous scientist, Finlay was quick to add: "These experiments are certainly favorable to my theory but I do not wish to exaggerate their value in considering them final . . . I understand but too well that nothing less than an absolutely incontrovertible demonstration will be required before the generality of my colleagues accept the theory so entirely at variance with the ideas which have until now prevailed about yellow-fever."[28] That "incontrovertible demonstration" would have to wait and would ultimately cost Jesse Lazear his life.

The response to Finlay's presentation to the academy that defied the conventional wisdom about yellow fever was predictable. According to Finlay's son, the lecture was greeted with "universal ridicule."[29] Despite the hostile reception, Finlay persisted. He advocated strongly for the destruction of mosquito larvae "in swamps, pools, privies, sinks, street-sewers and other stagnant waters," and he reiterated his belief that "the most essential point must be to prevent those insects from reaching yellow fever patients."[30] He also continued his investigations and freely shared his findings with those willing to listen. His subsequent inoculation experiments, however, produced few confirmed cases of yellow fever. (By 1894, he had reported only a handful of confirmed cases among more than one hundred experimental subjects.)[31] Unfortunately for the fight to eradicate yellow fever, Finlay's ideas were relegated to obscurity for years.

Finlay's findings languished, but in the meantime, new scientific evidence gradually began to accumulate that would eventually bolster his case. In 1897, the British physician Ronald Ross described the incubation and life cycle of the malaria parasite and showed that malaria was transmitted by the *Anopheles* mosquito.[32] Shortly thereafter, Henry Rose Carter, a US physician-researcher, discovered that when yellow fever arrived in a populated area, there was typically a gap of two weeks between the time the first isolated cases appeared and subsequent community spread. (This two-week incubation period readily explains why the vast majority of Finlay's inoculations, which were administered well short of the necessary incubation period, failed to produce cases of yellow fever.)

Why Carlos Finlay's pioneering research on the mosquito method of transmission in yellow fever was ignored for so long is an open question. Certainly, it contradicted his era's established medical wisdom about miasma or fomites as the culprit in spreading yellow fever. There was also Finlay's failure to convincingly demonstrate that his *Culex* mosquitoes reliably transmitted yellow fever from sick patients to previously uninfected volunteers. But other factors may also have been at work that explain why, even after Finlay's mosquito transmission theory was proven, he was not given due credit. Bias against Cubans and their national aspirations was likely the main reason Finlay didn't receive the credit he deserved. In an award-winning essay, the physician Daniel Liebowitz wrote, "Considering the U.S. climate of animosity towards Cuban resistance to American annexation, it is not surprising that Finlay's contribution . . . was largely diminished or entirely denied."[33] As evidence for his "political" interpretation, the author cites the belated decision by American officials to honor Finlay as a hero in the struggle to eradicate yellow fever at ceremonies in Philadelphia in 1952—at a time when a pro-American Cuban regime had just seized power. Finlay's discoveries—and his reputation—were held hostage for more than half a century by the state of US-Cuban relations. Our charge here is to shed light on issues of science, not politics, but it is worth noting that, as we have seen so clearly in the debate about COVID-19, political allegiances can easily inject themselves into discussions about medical evidence and force science to navigate a torturous path.

The Spanish-American War and the Creation of the Yellow Fever Board

With the battle cry "Remember the Maine!" still reverberating in their ears, US Army troops invaded Cuba on June 22, 1898. From the start, yellow fever, which had been endemic in Cuba since the mid-1600s, devastated troops on both sides of the conflict. An estimated sixteen thousand Spanish troops occupying the island had died from yellow fever between 1895 and 1898.[34] Within a month of the US invasion, the disease had begun to spread among the Fifth Army Corps. The coastal town of Siboney in southeastern Cuba, where US troops had landed, was the suspected source of the outbreak. With no understanding of how the disease was transmitted, General Nelson Miles ordered its residents evacuated and the town burned to the ground.

The fighting was over quickly, with the United States annexing Cuba in December 1898. Roughly five times as many American soldiers had died of yellow fever than from battlefield wounds.[35] It was clear that as part of its occupation of Cuba, the United States would have to contend with yellow fever. By 1893, George Sternberg, already considered "the father of American bacteriology," had been appointed US Army surgeon general. It fell to him to organize the mission to finally understand and get Cuba's endemic yellow fever under control. Sternberg, we may recall, had a history with the disease, having advanced his own bacterial theory of causation. In May 1900, Sternberg charged a commission with undertaking "scientific investigations with reference to the infectious diseases prevalent on the Island of Cuba."[36] Yellow fever was certainly Sternberg's primary target as it was the most virulent and lethal of Cuba's tropical illnesses and had no effective treatment.

To lead the Yellow Fever Board, Sternberg selected Walter Reed, a career US Army physician who had twice been posted to Baltimore and who on both occasions had trained at Johns Hopkins with William Welch. Reed was born in Belroi, Virginia, on September 13, 1851. His father, a traveling Methodist preacher, couldn't afford to have three sons attending college at the same time, so Walter was urged to enroll in the University of Virginia's shorter two-year medical program. He became the youngest graduate in the medical school's history.[37] After earning a second medical degree from Bellevue Hospital Medical College, he worked for several years at the New York City Board of Health, likely his first exposure to the emerging field of public health. Reed joined the US Army's Medical Corps in 1875. He moved more than fifteen times in his twenty-seven-year military career. While at Fort Apache in Arizona, Reed and his wife, Emilie, adopted a young Native American girl.

Reporting to Reed on the Yellow Fever Board were three other physicians. Jesse Lazear was born on May 2, 1866, in Baltimore. Lazear attended Washington and Jefferson College and then transferred to Johns Hopkins as an undergraduate. He went on to pursue a medical degree from Columbia University's College of Physicians and Surgeons. After graduating, Lazear traveled to Europe to study under August von Wassermann, the famed bacteriologist who succeeded Koch at the Institute for Infectious Diseases in Berlin, and Elie Metchnikoff, one of Europe's leading immunologists, at the Institut Pasteur in Paris. With his strong background in bacteriology, Lazear returned to the United States to head the clinical laboratory at Johns Hopkins Hospital (figure 6.2). In a letter to his wife, he explained he chose Hopkins because "there is no such group of men anywhere in the world. This is a great privilege to be able to work with such men."[38] At Hopkins, his training and interests prompted him to investigate malaria and yellow fever. In February 1900, Reed, probably acting on the recommendation of William Welch, chose Lazear to head the hospital laboratory at the Columbia Barracks in Cuba. Lazear was inducted into the army as a surgeon and resumed studying mosquitoes.

FIGURE 6.2. *Lazear as a young man, circa 1895.*

Joining Reed's Yellow Fever Board when it first met in Cuba in June 1900 were two other physicians (figure 6.3). James Carroll was an Englishman who had immigrated to Canada as a young adult. He had worked there as a lumberjack before moving to the United States and enlisting in the army. Carroll eventually developed an interest in medicine and earned an MD from the University of Maryland in 1891. While carrying out post-graduate work at Johns Hopkins under William Welch, Carroll met his future Yellow Fever Board colleagues Walter Reed and Jesse Lazear. All three young men could rightly be called Welch disciples. The final member of the Yellow Fever Board was Aristides Agramonte y Simoni. Born in Cuba, he immigrated with his mother to New York after his father, a revolutionary, was killed in the first Cuban War of Independence. Agramonte attended medical school at Columbia University, where he graduated with honors. One of his classmates was Jesse Lazear. Reed and his Yellow Fever Board colleagues had overlapping professional histories, and three of the four had trained at Hopkins. Their shared experiences would prove fruitful. In Lazear's case, his interest and experience in handling mosquitoes would be fateful.[39]

FIGURE 6.3. *Aristides Agramonte, Jesse Lazear, and James Carroll standing outside the Columbia Barracks in Cuba in 1900. Both Carroll and Lazear would volunteer to be bitten by an infected mosquito, but only Carroll would survive the infection.*

The Work Begins in Earnest

The team that officially launched its investigations in Cuba in June 1900 took a two-pronged approach to uncovering the etiology of yellow fever. As noted, Surgeon General Sternberg was eager to disprove Sanarelli's theory of *Bacillus icteroides* as the cause of the disease. This proved relatively easy. In July 1900, about a month after the Yellow Fever Board started its investigations, Lazear and Carroll, working in the bacteriology laboratory at Columbia Barracks, took cultures of blood, internal organs, and body cavities from twenty-nine patients with yellow fever, both living and dead. Not a single culture grew *Bacillus icteroides*. The conclusion was inescapable: Sanarelli's bacterium was not the pathogen that caused yellow fever. Nor, in fact, was Sternberg's *Bacillus X*.

In the meantime, a severe outbreak of yellow fever at the Columbia Barracks in the summer of 1900 added urgency to the second prong of the Yellow Fever Board's work—understanding how the often-lethal disease was transmitted. Two major possibilities were explored. On the one hand, Reed's team considered person-to-person spread by means of bodily secretions. On the other, the possibility that mosquitoes might be the vector that transmitted the disease was beginning to gain traction with at least some members of Reed's team.

Several events converged to spur the Yellow Fever Board's interest in the mosquito theory of transmission, the same compelling but unproven explanation advanced decades earlier by Carlos Finlay but largely relegated to the sidelines. On July 17, Lazear met with Henry Rose Carter. Carter, as mentioned earlier, was a doctor in the US Marine Hospital Service, a forerunner of today's US Public Health Service. He was assigned to Havana as chief quarantine officer since Cuba was presumed to be the source of many yellow fever outbreaks in the United States. In 1898, the year before Carter was transferred to Cuba, he had carefully observed the interval between the first appearance of isolated cases of yellow fever and subsequent outbreaks in two small Mississippi towns.[40] Carter determined that the agent that caused yellow fever "must undergo some change in the environment before it is capable of infecting another man."[41] Calling this interval the "period of extrinsic incubation," he set the duration at about two weeks.[42] At their meeting, Carter described to Lazear his work in Mississippi on the timing of the transmission of the disease. The next day he sent Lazear a copy of the published article on his investigations with a cover note that read in part, "As I said, to me the a-priori argument for Dr. F's. [Finlay's] theory has much more in its favor and to me is more than plausible, although his observations as I read them are not convincing, scarcely corroborative."[43] It could hardly have escaped Lazear's attention that Carter's findings about the extrinsic incubation period went a long way toward explaining Finlay's failure to produce a higher number of yellow fever cases among experimental subjects inoculated using infected mosquitoes.

On or about July 18, Herbert E. Durham and Walter Myers from the prestigious Liverpool School of Tropical Medicine arrived in Havana en route to study yellow fever in Brazil. Their conversations with Reed, Carroll, and Lazear, as well as with Carter and Finlay himself, served to deepen the American scientists' interest in mosquito transmission and lent additional credibility to Finlay's earlier work.[44] The two British scientists seemed to have been mostly concerned with Carter's hypothesis about an incubation period. They appeared satisfied that Carter's two-week incubation period accounted for the timing of yellow fever spread. Notably, Durham and Myers commented on recent discoveries about the transmission of malaria. In light of these discoveries and Carter's proposed incubation period, Finlay's suggestion twenty years earlier that the disease was spread by mosquitoes "hardly appears so fanciful" as "when the idea was first broached."[45] In the encounter with the British scientists, as at other stages in the initial investigation of yellow fever, Lazear seemed more interested in the mosquito theory than his Yellow Fever Board colleagues were.[46]

The final event that prodded the Yellow Fever Board's interest in the mosquito theory of transmission was an encounter with its originator. Carlos Finlay, whose ideas had been ridiculed decades earlier by his fellow scientists, lived at Aguacate Street No. 110 in Havana, just miles from the Yellow Fever Board's headquarters at Columbia Barracks. It is not clear who from the board sought out Finlay's advice or precisely when, but it seems plausible that it was Lazear, especially in light of Lazear's June 26, 1900, meeting with Carter, in which the Marine Hospital physician shared his growing confidence in Finlay's mosquito transmission theory. While in Baltimore, Lazear had investigated the role of mosquitoes in the transmission of malaria. After arriving in Cuba but before taking his place on the board, Lazear had captured mosquitoes in Quemados, the site of a major yellow fever outbreak, in a failed effort to dissect and examine them microscopically for the yellow fever pathogen. (We noted earlier that the flavivirus that causes yellow fever is too small to be seen using a standard microscope.) Lazear and at least one of his colleagues likely visited Finlay more than once. It is hard not to imagine Finlay's enthusiasm at finally being consulted by leading scientists. We know for certain Finlay provided his visitors from the United States with a porcelain dish containing eggs from the *Aedes aegypti* mosquito.[47]

The Beginning of the End

When it came to the unsolved question of transmission, Reed and his team knew of no animal model for testing the competing hypotheses of spread via bodily secretions or the bite of mosquitoes. Accordingly, they decided to use human subjects in their investigations. Carroll, taking note of a July 1900 meeting with his colleagues, wrote: "The serious nature of the work [was] decided upon and the risks . . . were fully considered, and we all agreed as the best justification we could offer for experimentation upon others, [was] to submit to the same risk of inoculation ourselves."[48] The decision by the board to experiment on themselves was reached by a vote. It's worth noting that, by today's standards, self-experimentation, though definitely setting a higher bar ethically than simply using enlisted subjects, is not at all the same as seeking informed consent.

The Yellow Fever Board's work to investigate the method of transmission unfolded in two distinct stages. As the board's most experienced bacteriologist and someone who had studied mosquitoes in connection with malaria, Lazear spearheaded an initial effort to confirm Finlay's mosquito theory of transmission (figure 6.4). He laboriously hatched the *Aedes aegypti* eggs Finlay had given him. Lazear dubbed the adult female mosquitoes his "birds" and kept them in separate test tubes. In *Yellow Jack*, Pierce and Writer give this dramatic account of what Lazear did next:

> At Las Animas Hospital [the Cuban hospital where most of the American soldiers sick with yellow fever were being treated], he would allow [the mosquitoes] to feed on yellow fever patients in the active stage of the disease; this became known as "loading." He would load them by taking the test tube each was in, turning it upside down, and gently tapping the tube to encourage the mosquito to move upward. When the mosquito was at the top of the inverted test tube, he would remove the cotton plug, and place the open mouth of the test tube against the patient's skin. Then he would wait for the bird to light on the patient and fill itself with blood. Once filled, the mosquito would instinctively become airborne and Lazear would quickly replace the cotton plug. Notes were made as to the patient or patients each bird fed on and what day of illness they were in at the time.[49]

Lazear carried out nine "inoculations" between August 11 and August 25 by allowing his birds to bite eight healthy volunteers, including himself.[50] He did not get sick, nor did any of the seven other volunteers he inoculated. Despite his initial belief in mosquito transmission, Lazear was skeptical about Carter's evidence of a twelve-day incubation period before the mosquitoes could spread the disease.

Lazear's next volunteer was his friend and colleague James Carroll. The circumstances were almost accidental. On August 27, not at all sure his mosquitoes were able to transmit yellow fever, Lazear departed from his normal routine and transported the birds from Las Animas back to Columbia Barracks. He and Carroll were working in the laboratory that afternoon. Their conversation turned to the mosquitoes. Carroll had serious reservations about the mosquito theory. When Lazear mentioned that one of the birds had failed to feed on a patient's blood earlier that day and might die, Carroll, more or less as a lark, volunteered to let the insect bite him.[51] This mosquito had bitten an infected patient named Mulcahey twelve days earlier, on the second day of the soldier's illness, according to the large notebook Lazear kept. Lazear took the test tube holding the languishing mosquito, inverted it in on Carroll's arm, and coaxed the bird to feed. He was not at all convinced that the failing mosquito would sicken Carroll. In a letter to his mother written later that day, Lazear didn't even mention his latest experiment.

FIGURE 6.4. *Jesse Lazear within months of taking up his post with the Yellow Fever Board in Cuba.*

Carroll was treating patients at Las Animas on August 29 when he started to feel ill. He performed a blood test on himself for malaria. The result was negative. A day later, Carroll's temperature had reached 102 degrees.[52] Another blood test for malaria, this time administered by Lazear, was negative. On September 1, Carol was transferred to the yellow fever ward at Columbia Barracks, where his condition continued to worsen. By the next day, there was albumin in his urine and his temperature reached 103.6.[53] On September 3, he had the telltale jaundiced skin color of a yellow fever victim. Carroll was clearly in danger of becoming a casualty of Lazear's research.

Lazear must have felt terrible fear and guilt at having caused the life-threatening illness of a friend and professional colleague, especially one who had a wife and five children. Agramonte wrote, "Lazear and I were almost panic-stricken. Lazear, poor fellow, in his desire to exculpate himself . . . repeatedly mentioned the fact that he himself had been bitten two weeks before."[54] At the same time, Carroll's illness was indisputably a strong indication that Finlay's long-contested theory of mosquito transmission was correct, and thus the first tangible evidence that his painstaking work with *Aedes aegypti* had finally borne fruit.

Though it strongly suggested that Finlay had been right, Carroll's case wasn't scientifically conclusive evidence for the mosquito theory. He had not isolated himself as the board's experimental protocols required. In fact, Carroll had multiple direct contacts with yellow fever patients before being bitten by the experimental mosquito. The origin of his yellow fever was thus suspect. He could have been infected by any of these contacts. In need of a volunteer with no prior exposure to the disease, Lazear recruited Private William H. Dean. A recent arrival in Cuba, Dean had not had any contacts with patients with yellow fever.

On August 31, the same day Carroll became bedridden, Agramonte asked Dean if he would participate in Lazear's mosquito experiment. Dean agreed. Lazear used the same mosquito that had bitten Carroll as well as three others to inoculate Dean that afternoon. Five days later, Dean became ill.[55] At first he was merely tired and a bit dizzy, and then Dean's illness progressed rapidly through the classic yellow fever symptoms from a high fever and bloodshot eyes to albumin in his urine and eventually jaundice.[56] Fortunately, Dean did not develop the black vomit that usually presages death. The enlistee was much younger than Carroll, and in the end his case was less severe. A few days after his symptoms peaked, Dean was on the mend. Carroll, too, despite his more severe infection, was recovering by September 8.

Perhaps relieved that two patients he had experimentally infected were out of danger, Lazear allowed himself a moment of quiet celebration about his hard-won discovery. On September 8, he wrote his wife, "I rather think I am on the track of the real germ, but nothing must be said as yet, not even a hint. I have not mentioned it to a soul."[57] Among those Lazear kept in the dark about his mosquito experiments was his commanding officer Walter Reed, who was in the United States on other duties and by this time on vacation in the mountains of Pennsylvania.[58] Lazear's failure to inform Reed about his experiments sent events careening on an unforeseen course.

Lazear had already allowed himself to be bitten twice by mosquitoes that had recently fed on yellow fever patients. Both times he remained healthy. It's not altogether surprising, however, that he would try his self-experiment again in an effort to prove Finlay's theory of transmission. Lazear's interest in testing out the mosquito theory using himself as a guinea pig may well have increased after Carroll and Dean both developed full-blown yellow fever after being bitten. It is also worth noting that Lazear's dangerous experiments on himself were in keeping with the Yellow Fever Board's earlier decision to submit themselves to the same risks that they were asking others to take on.

As a scientist, Lazear recognized that only Dean's case would stand up to rigorous scrutiny. On September 13, 1900, Lazear was working at Las Animas Hospital to load his mosquitoes by allowing them to feed on the blood of yellow fever patients when a stray mosquito landed on the back of his hand. He made no effort to brush the mosquito away because he didn't think it was a specimen of the *Aedes aegypti* species that spread the disease.[59] At least this was the story that Lazear told Carroll, who was recovering from his own experimental case.[60] Most of Lazear's colleagues seem to have accepted this account at face value. Certainly, when the Yellow Fever Board issued its report in October 1900, it reiterated Lazear's description of the incident.[61]

Was this what happened? There are strong indications that events unfolded quite differently, and that Lazear deliberately infected himself. Lazear kept a notebook that documented the team's clinical and laboratory observations. The entry on page 100, written in Lazear's hand, reads in part as follows:

> Guinea pig No. 1—red
> Sep. 13 This guinea pig bitten today by a mosquito which developed from egg laid by a mosquito which bit Tanner—8/6. This mosquito bit Suarez 8/30.[62]

Several facts point toward the conclusion that Lazear's notebook entry is a more accurate description of events, that he was "Guinea pig No. 1," and that he intentionally inoculated himself with a mosquito he knew carried the yellow fever pathogen.[63] The identical date, September 13, in Lazear's account to Carroll of when he was bitten *and* in his notebook is too big a coincidence to ignore. What's more, Lazear was far too experienced with mosquitoes, from his work both on malaria and yellow fever, to not have recognized the distinctive markings and body shape of the *Aedes aegypti* if one had in fact landed on the back of his hand. Lazear had raised the mosquitoes from eggs in the laboratory and even taught his colleagues how to distinguish the disease-carrying females from noninfectious males.

Personal Tragedy and Final Conquest

On September 18, five days after he was bitten by the infected mosquito, Lazear stayed in bed and missed work. A member of the camp's medical staff diagnosed his illness as yellow fever. His condition deteriorated rapidly. He was moved to the yellow fever wards at Columbia Barracks hospital, but such treatment as existed at the time was to no avail. The jagged black line on Jesse Lazear's temperature chart dropped precipitously on September 24 and ends abruptly after the evening reading on September 25 with the blunt but clinically precise notation "Died 8:45 P.M." (figure 6.1). It fell to Major Jefferson Randolph Kean, attached to the Seventh Army Corps and a great-grandson of Thomas Jefferson, to notify Lazear's wife. Kean apparently had not been informed that Mabel Lazear knew nothing about her husband's grave illness. The Western Union telegram arriving at the Beverly, Massachusetts, home where she was staying after having given birth just two weeks earlier must have landed with devastating effect. It simply read, "Dr. Lazear died at 8 this evening. Kean" (figure 6.5).

Walter Reed was wrapping up his stay in the United States when Lazear first became ill. On September 24, a letter from Carroll about Lazear's worsening condition arrived.[64] When he learned of Lazear's experiments and their dire consequences, Reed immediately replied to Carroll, writing of his grave concern about Lazear and expressing the hope that his relative youth and sturdy constitution would spare him. At some point, however,

Reed's compassion as a friend and colleague must have collided with his training as a scientist. He recognized that neither Lazear's case nor Carroll's would be helpful in proving the mosquito theory of transmission. He chided Carroll:

> If you, my dear Doctor, had, prior to your bite remained at Camp Columbia for ten days, then we would have *a clear case, but you didn't!* You went just where you might have contracted the disease from another source . . . Unfortunately Lazear was bitten at Las Animas Hospital! *That* knocks his case out; I mean as a thoroughly scientific experiment . . . I am only regretting that two such valuable lives have been put in jeopardy, under circumstances in which the results . . . would not be above criticism.[65]

Reed ended his reply to Carroll by noting "his" belief in mosquito transmission (though the evidence suggests that Reed was a late convert to the theory), but he was clearly upset by the dangerous experiments used to test the idea.

Jesse Lazear was buried with full military honors on Monday afternoon, September 26, 1900, in a temporary military cemetery a short distance from Columbia Barracks. Lazear's remains were returned to his birth city eighteen months later. They now rest in a family plot in Baltimore's Loudon Park Cemetery (figure 6.6). His courageous death set the stage for a final assault on yellow fever, a disease that had held Cuba and the Americas hostage for nearly three centuries.

Walter Reed returned to Cuba in October 1900.[66] He informed General Leonard Wood, the army's commanding officer there, that the Yellow Fever Board had a newly focused mission: first, to prove once and for all that the *Aedes aegypti* mosquito was the host organism and agent of transmission for yellow fever, and, second, to show definitively that the fomite theory of yellow fever transmission was false and that, as a result, close contact with clothing and other belongings from patients with yellow fever could not spread the disease. Shortly after his return, Reed negotiated the purchase of a small plot of land about a mile from Columbia Barracks. The area was sufficiently remote that it could be used to test these hypotheses in carefully controlled experiments. This last element was crucial and a defining characteristic of the scientific method that Reed and his Yellow Fever Board applied to their efforts to understand and eradicate the disease. Unlike Lazear, who had died in a heroic but uncontrolled experiment, in each of the board's tests to identify possible causative agents, some of the volunteer subjects were exposed to mosquitoes and the others were carefully isolated from mosquitoes.

FIGURE 6.5. *Jesse Lazear, with his infant son, William Houston Lazear, in Cuba. Lazear would suffer a painful death from his self-experimentation within months.*

FIGURE 6.6. *Coauthor Ralph Hruban visiting Lazear's grave at Loudon Park Cemetery.*

Reed named the experimental area Camp Lazear in honor of his fallen colleague. Small wooden buildings were erected, and volunteers were recruited from among the soldiers at Columbia Barracks and recently arrived immigrants from Spain (figure 6.7). Several elements of informed consent were used. Reed developed a printed consent form, which was translated into Spanish, that acknowledged the risks and outlined the volunteer subjects' rights. These rights included access to medical care and compensation—in this case $100 in gold for participating in the experiments and an additional $100 if they became sick.[67] Certainly, times have changed. Today, we might question whether paying an impoverished person $100 to risk their life is truly offering them informed consent.

Under the experimental protocol, Building 1, the "Infected Clothing Building," was filled with sheets, blankets, pillowcases, and other supplies used by yellow fever patients and "purposely soiled with a liberal quantity of black vomit, urine and fecal matter."[68] Volunteers assigned to Building 1 were even required to wear the soiled clothing of people who had died from yellow fever. The building was fully screened to prevent any exposure to mosquitoes. Volunteers stayed in these quarters for twenty days. A recent medical journal article leaves little doubt about the intimate contact volunteers testing the fomite theory had with contaminated materials. The authors write, the "small frame-building . . . had been packed with the bedding of yellow fever patients that was heavily soiled with urine, faecal matter and the characteristic black vomit . . . The building was occupied at night by two or three non-immunes, who slept in the contaminated bedding, and in one case in the nightshirts of the patients."[69] It could not have been pleasant. Another account said that the odor in Building 1 was so foul that the test subjects fled only to return later that night to sleep in their soiled beds.[70] None of the volunteers exposed to bedding and clothing soaked with excretions from patients with yellow fever contracted the disease.

FIGURE 6.7. *The Fomite Building at Camp Lazear. Volunteers would spend days in small rooms filled with bedding soaked with the bodily excretions from patients with yellow fever.*

About seventy-five yards to the south was Building 2, the "Infected Mosquito Building." It was divided in half by a wire screen, with one bed on the mosquito-bite side and two more on the non-mosquito, or control, side. In his biography of his Hopkins colleague Walter Reed, Howard A. Kelly, one of the School of Medicine's four founding doctors, gave a detailed account of the exposures the volunteers underwent in Building 2. Kelly wrote that five "promising" mosquitoes were selected by the experimental team on December 5, 1900, and introduced into the exposure side of the mosquito building. The five insects, he noted:

> Had been contaminated fifteen days previously; one, nineteen days; and two, twenty-two days. These were allowed, with his free consent, to bite the patient, John R. Kissinger; the result of the bites was perfectly successful, for about midnight on December 8, at the end of three days, nine and a half hours, the subject, who had been under strict quarantine during fifteen days, was seized with a chill that proved the beginning of a well-marked attack of yellow fever.[71]

Other volunteers who were bitten by the infected mosquitoes also became sick. The volunteers on the control side of the mosquito building remained healthy. Eighteen volunteers, including Lazear and Carroll, all of whom had been exposed to mosquitoes, were experimentally infected by February 1901.[72] Four of the eighteen, including Lazear, died of yellow fever.[73]

The carefully controlled experiments provided Reed and the board with the rigorous evidence on the cause and method of transmission of a deadly disease that had remained elusive for two and a half centuries. The Yellow Fever Board's controlled experiments were groundbreaking in another respect. They helped embed once and for all the principle of informed consent in American medicine. In the handful of years between Sanarelli's experiments attempting to prove a supposed bacterial cause of the disease and the work at Camp Lazear, we can trace a steady progression in this fundamental concept of scientific medicine. Sanarelli didn't ask his patients to volunteer, and he apparently injected them with dangerous bacteria without their knowledge or consent. Lazear and his colleagues agreed that such a practice was unethical and that the only justification for experimenting on other human beings was a willingness to submit to the same procedures themselves. That willingness cost Lazear his life. In the final effort to prove the mosquito theory of transmission, Reed and the board used written consent, compensation, and guaranteed medical treatment, practices that are standard in medical experimentation on human subjects today.

Once the mosquito's role in transmitting yellow fever was proven, an all-out campaign to eradicate it began. In October 1900, Jefferson Randolph Kean ordered military personnel in western Cuba to eliminate standing water, the mosquito's hatching ground. William Gorgas, the chief sanitation officer for Havana, although initially a mosquito skeptic, began an anti-mosquito campaign that included suffocating the developing larvae by spreading a thin layer of oil or kerosene on the surface of the pools of standing water. By December, the order had been extended to cover standing water on all army bases in Cuba. In February 1901, Gorgas's squads of mosquito inspectors launched an all-out anti-mosquito program in Havana, where decades earlier Carlos Finlay had proposed a similar eradication campaign to no avail.[74] The results were dramatic. After five months of Gorgas's eradication program, there were no further cases of yellow fever in Havana.[75]

Walter Reed presented his paper "The Etiology of Yellow Fever—An Additional Note" at the February 6, 1901, meeting of the Pan-American Medical Association. Reed and his Yellow Fever Board colleagues James Carroll and Aristides Agramonte were listed as authors. The paper paid special tribute to Jesse Lazear. It summarized the results of the experimental work carried out at Camp Lazear and laid out eleven conclusions.[76] These ranged from the *Aedes aegypti* mosquito's role as the intermediary host of the yellow fever organism and the necessary incubation period before the mosquito's bite can transmit the disease, to the definitive finding that yellow fever cannot be spread by contact with a sick person's bodily secretions or their clothing or other possessions. In well over a century of further research, none of these conclusions has been overturned.

The Courage and Devotion of the Soldier

On October 6, 1900, Reed wrote his wife, Emilie, that Lazear "was a splendid, brave fellow, and I lament his loss more than words can tell; but his death was not in vain—His name will live in the history of those who have beautified humanity."[77] Later, on December 9, 1900, with what was a mix of elation and perhaps relief that the first phase of the struggle against yellow fever was over, Reed wrote to her:

> Rejoice with me, sweetheart, as, aside from the antitoxin of diphtheria and Koch's discovery of the tubercle bacillus, [the discovery of the mosquito's role in spreading yellow fever] will be regarded as the most important piece of work, scientifically, during the 19th century. I do not exaggerate, and I could shout for very joy that heaven has permitted me to establish this wonderful way of propagating yellow fever.[78]

The search for the as-yet-undiscovered virus that was the ultimate cause of yellow fever and the cure (in the form of a vaccine) would continue for decades. The immense contribution to the fight against yellow fever made by Finlay, Lazear, Reed, and their colleagues is nonetheless indisputable.

Reed went on to direct the Surgeon General's Library, the same institution that John Shaw Billings had helped grow. Tragically, he would live for less than two years after his historic report to the Pan-American Medical Association. He died of acute appendicitis on November 23, 1902. His lifetime of outstanding national service and medical achievement is memorialized at Walter Reed Medical Center in Washington, DC, named in his honor.

As for Jesse Lazear, he, too, has been publicly remembered. When visitors enter the National Cathedral in Washington, DC, and gaze up at the stained glass, they can see the beautiful window titled *Sacrifice* (color plate 10). Many people don't understand what they are looking at, but perhaps readers of this chapter will. In the panels honoring Lazear, we see his laboratory workbench filled with beakers and flasks, but also the lethal mosquito that this brave young scientist allowed to bite him. At

Johns Hopkins Hospital where one of us (Ralph Hruban) works, there is a plaque honoring Lazear in one of the main corridors. After detailing his education and training, it notes that he was a member of the Yellow Fever Board in Cuba in 1900 and that he died fighting the disease in September of that year. The final words on the plaque read: "With more than the courage and devotion of the soldier he risked and lost his life to show how a fearful pestilence is communicated and how its ravages may be prevented" (figure 6.8). Lazear's life, given in the name of science, was the ultimate sacrifice.

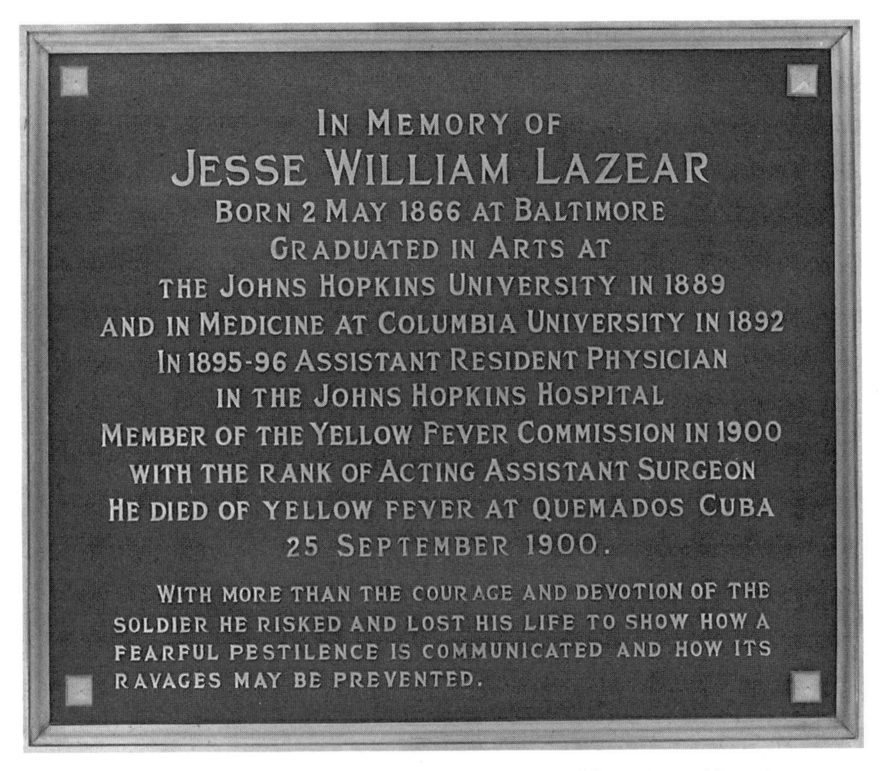

FIGURE 6.8. *Plaque honoring Jesse Lazear in one of the main corridors of the Johns Hopkins Hospital. (Photograph by Norman Barker.)*

When we think about Lazear, we cannot help being reminded, as we were in the case of Halsted, of exceptional courage and extraordinary devotion to the progress of scientific medicine. Both men, in very different circumstances, were willing to experiment on themselves and made a conscious decision to put saving human lives above personal risk. In our own time, shaped as it has been by a catastrophic pandemic and untold suffering and loss of human life, it is hard not to also think about the men and women who have fought on the front lines treating patients with COVID-19. The disease they have battled may be different and today's social and technological environment radically distinct, but the dedication and self-sacrifice of these healthcare workers vividly reminds us of the price paid by William Stewart Halsted and Jesse Lazear.

7

MAX BRÖDEL

Art Applied to Medicine

Regarded by many in his field as the most brilliant anatomical artist since Leonardo da Vinci and generally credited with developing medical illustration as a professional discipline, Max Brödel exemplifies the pivotal role nonphysicians have played in the development of scientific medicine. Brödel's artwork, striking both for its beauty and its anatomical insight, carries "graphic bolts of melody as it . . . so often stuns and dazzles."[1] It almost invariably evokes what biographer and fellow medical illustrator

John Cody has described as a "wow reaction . . . the natural response to a demonstration of any sort of extraordinary human giftedness."[2] Yet, to fully appreciate Brödel's pivotal contributions to medicine, we need to look back to the distant past.

Art and Science: A Partnership That Predates Written History

Throughout human history, art and science have been allies in the pursuit of knowledge. Artists needed to understand human anatomy to create accurate depictions of the human form, and scientists relied on artists to communicate their findings about the body's structure and function. The partnership goes back many millennia. Cave drawings from the Pyrenees region of France dating from more than thirty thousand years ago depict arrows embedded in the hearts of the prehistoric bison that early inhabitants of the area hunted.[3] This Paleolithic art, though not scientific, betrays the first glimmer of human understanding about the central role of the heart in sustaining life.

In the second century C.E., the physician and philosopher Claudius Galenus (Galen of Pergamon) laid the foundations of anatomy and physiology, mostly by dissecting monkeys since dissections of human cadavers were strictly forbidden.[4] Galen's drawings were often flawed or fanciful. He made obvious mistakes by assuming that simian and human anatomy were identical. For centuries and well into medieval times, Galen's authority was so great that few scholars were willing to question it.[5] During the Renaissance, Leonardo da Vinci applied his extraordinary talents as a draftsman and his probing curiosity about nature to create remarkably accurate and detailed anatomical drawings of the human body. He dissected more than thirty human cadavers to deepen his knowledge of human anatomy and his skills in portraying human forms. Still, Leonardo's drawings remained private and unpublished for centuries.

Not until Andreas Vesalius, the Brussels-born physician and chair of surgery at the University of Padua, published his *De humani corporis fabrica libri septem* (On the Fabric of the Human Body in Seven Books) in 1543 was the flawed anatomy of Galen put under public scrutiny.[6] Like Leonardo, Vesalius identified errors Galen had made, but unlike Leonardo, Vesalius openly questioned Galen in his first-of-its-kind anatomy text, writing that the Roman physician had been "fooled by his monkeys."[7] He relied on dissections of human cadavers and employed Jan Stephan van Calcar, a disciple of the famed Renaissance painter Titian, and other skilled artists to create the work's elaborately illustrated plates. The book was a milestone in the science of anatomy and in the productive partnership between art and medicine. Once Vesalius had demonstrated the importance of questioning the dogma of the past and the power of accurately realized drawings in medical texts, there was no going back.

Still, the world would have to wait another 350 years before Max Brödel set the discipline of medical illustration on a solid and sustainable course. In the meantime, as one expert on the history of medical illustration has pointed out, both artists and physicians struggled in their own realms.[8] Physicians were hard-pressed to find artists with the highly specialized skills needed to create a visual record of their new discoveries. Lacking these collaborators, doctors often had to resort to their own crude drawings. By the same token, few artists were willing to undertake the lengthy, malodorous dissections (especially in an era before refrigeration and preservatives) needed to create faithful images of the human body.[9]

Brödel changed not only the history but the very conception of medical illustration. He combined superb artistic skills, an approach rooted in science, and rare gifts as a collaborator and a teacher. Together, these qualities made it possible for him to dramatically expand the scope and caliber of anatomical illustration and its usefulness in medical research and practice. All these accomplishments lay in the future on the day a blue-eyed, curly-haired young German immigrant, speaking barely a word of English, landed in Baltimore's harbor, sporting a dapper black derby.[10] In this chapter, we explore Brödel's life story in its public and private dimensions. We also look at the hallmarks of his genius as a medical artist, his impact on Hopkins's growing renown in medicine, and, finally, his legacy in medical illustration.

Talent That Blossomed Early

Paul Heinrich Max Brödel was born in Leipzig, Germany on June 8, 1870. His father, Paul Heinrich Louis Brödel, with whom Max would come to share artistic interests, made musical instruments and worked for the Steinweg Piano Company. Young Max was musically talented, and even though his father forced him to practice the piano endlessly, Max developed a lifelong passion for the instrument.[11] By the age of eight, his piano teacher set Max to work on the oversized challenge of Beethoven's *Sonata Appassionata*. Many years later, Brödel told a friend, he was deeply moved by "its beauty and stirring qualities."[12] Brödel's early flashes of artistic talent weren't limited to music. He soon developed an interest—and a gift—for drawing and painting. Around this time, Brödel contracted smallpox. Max seemed to have nine lives, as this was just the first of many life-threatening illnesses he would survive.

In his teens, despite his father's insistence that Max make his concert debut and take up a musical career, Brödel was permitted to pursue his interest in art. He attended the Leipzig art school (Die Königliche Kunstakademie und Kunstgewerbeschule zu Leipzig) between 1885 and 1890. There Brödel acquired what biographer Ranice Crosby would describe as "that delicate, loose touch that forever after graced whatever he did."[13]

In the summer of 1888, with Brödel now an adult and needing money to continue his studies, his artistic career took a crucial turn. With no scientific or medical training, he went to work for Dr. Carl Ludwig during the art school's recess at the University of Leipzig's renowned Institute of Physiology. At the time, Ludwig was one of the most prominent scientists in the world. His lab, as Brödel described it, was "the Mecca for medical men of all classes and all countries"[14] (figure 7.1). Ludwig asked Brödel to draw a magnified section of brain cortex. Brödel recalled the experience this way: "I worked many weeks to make that one drawing, my first medical picture, and I believe the hardest I ever attempted."[15]

Brödel's time in Ludwig's laboratory shaped his career in important ways. He gained unique insights into medical illustration. Perhaps the key insight was that creating a high-quality drawing wasn't simply a matter of faithful replication: "I blundered when I relied on faithful copying alone. Copying a medical object is not medical illustrating. The camera copies as well, and often better, than the eye and hand . . . but in medical drawing full comprehension must precede execution."[16] It was also in Ludwig's lab that Brödel met William Welch, who would go on to become a professor of pathology and a founding member of the Johns Hopkins School of Medicine faculty. There too he was introduced to Franklin Mall, the prominent anatomist, who some six years later would recruit Brödel to Baltimore. Welch and Mall were both visiting from America. As Brödel wrote, "It was also lucky for me to be poor, for I had to seek work during the summer vacations and other free hours throughout the year. I came under the eye of Prof. Carl Ludwig, the great physiologist, and was permitted to illustrate his research and that of his famous pupils. In the course of this work I met Dr. F. P. Mall and Dr. William H. Welch."[17]

FIGURE 7.1. *Max Brödel's pen-and-ink drawing of the Austrian physiologist Karl Ludwig. Ludwig was one of the greatest scientists of his time. Brödel would meet William Welch and Franklin Mall while working in Ludwig's laboratory.*

As with most able-bodied young men in the newly unified Germany, Brödel was drafted into the army and served for two years. His artistic talent spared Brödel some of the rigors of military service. Brödel spent at least a portion of his time creating drawings for officers and their families.

Because the illustrations were for his superiors, he couldn't charge for them, but he doubtlessly was treated better than many other enlisted soldiers.

After serving in the army, Brödel returned to Ludwig's lab. The acquaintance with eminent physician-scientists from America forged there earlier now bore fruit. Franklin Mall, recently named the first professor of anatomy at Hopkins, wrote to Brödel on September 8, 1893: "My dear Herr Brödel . . . I hope it will be possible to get you to Baltimore."[18] Accepting Mall's invitation, Brödel sailed for his new home on the steamship *Dresden* and arrived in Baltimore on January 18, 1894 (figure 7.2).

FIGURE 7.2. *Brödel in 1894, the year the young German immigrant arrived in Baltimore.*

As often happens with new immigrants, Brödel's first hours in his adopted country didn't go altogether smoothly. On his arrival in Baltimore, Brödel almost missed Mall, as there was some confusion about precisely where in Baltimore's expansive waterfront they were supposed to rendezvous. Both men boarded local ferries that sailed past each other in the harbor. Ranice Crosby writes: "Max noticed that one man [on a passing ferry] began gesticulating more vigorously than the rest, frantically pointing first ahead toward Locust Point and then back toward the terminal he had just left . . . Max squinted . . . and realized the man it was Dr. Mall."[19]

In the Right Place at the Right Time

There was another unanticipated twist before Brödel could settle into his new role at Hopkins. By the time he arrived in Baltimore, Mall's research had taken a different direction. He no longer needed an illustrator, so Brödel began his career by creating illustrations for Howard Kelly. Kelly was the first chair of the Department of Gynecology and Obstetrics, who at the time of Brödel's arrival was working on his magnum opus *Operative Gynecology*. One of the founding physicians of the School of Medicine depicted in *The Four Doctors* (see color plate 2, where he is seated on the far right), Kelly was a gifted surgeon and is recognized as one of the pioneers of modern gynecological surgery.

Brödel was ambitious and willing to work hard. As he wryly put it, "I will try to grab the lucky pig by its tail and then the chase begins, through thick and thin."[20] He was also in the right place at the right time. His friend and colleague, gynecologist Thomas Cullen, later wrote, "The time was ripe for Max. He came just as medicine and surgery were making greater strides than they had done in centuries and when many new illustrations were necessary."[21]

To master anatomy and pathology, Brödel tirelessly observed surgery and dissected cadavers in the autopsy room (figure 7.3). William Welch, newly appointed as the first chief of pathology at Hopkins, issued instructions that Brödel be given unfettered access to the autopsy rooms at the hospital.[22] As Brödel wrote, "The artist must first fully comprehend the subject matter from every standpoint: anatomical, topographical, histological, pathological, medical and surgical."[23]

FIGURE 7.3. *Brödel prepared for his illustrations by dissecting corpses and conducting careful microscopic studies.*

Brödel's future wife, Ruth Huntington Brödel, described Max's approach to creating a medical illustration this way: "Long hours of painstaking studies . . . laid the foundation of a masterful knowledge of anatomy. He was a keen observer of every detail."[24] Brödel's scientific approach was particularly remarkable for someone without any university training.

In Brödel's hands, dissection was a tool for unearthing new knowledge, not just communicating existing information. Just as Vesalius had found flaws in Galen's anatomy when he dissected cadavers in the sixteenth century, Brödel insisted that "while making a drawing the conscientious artist has a way of discovering gaps in contemporary knowledge."[25] At another point, Brödel declared, "the making of a sketch or a model produced new ideas of benefit to the investigation and . . . aided in the discovery of incorrect conceptions and . . . pointed out the best method to overcome the difficulty."[26]

In Brödel's view, the medical artist's work could not only reveal missing anatomical and pathological information, but also show new ways to overcome surgical obstacles. A case in point is the so-called Brödel line. After conducting exhaustive anatomical investigations, Brödel revealed that by cutting along a particular edge of the kidney, surgeons could avoid major blood vessels. His discovery is still in use in kidney surgery today.

Brödel's focus in his early years at Johns Hopkins was on illustrating Howard Kelly's encyclopedic work on surgical gynecology. Thomas Cullen, who was Kelly's chief resident and a lifelong friend of Brödel's, noted, "When Dr. Kelly's two volumes on *Operative Gynecology* appeared in 1898 Dr. Kelly was at once recognized as the leader of American gynecology, and Brödel's illustrations in these two volumes immediately revolutionized medical illustrating."[27]

Seeing the Unseen: Hallmarks of Brödel's Genius

Take a look at Brödel's illustrations, for example figure 7.4, a cutaway view of a breech delivery, and figure 7.5, an operation on the pituitary gland. Even the untrained observer is astonished by these masterpieces. Bodily tissues glisten realistically. Organs appear almost moist to the touch. Light and shadow fall in just the right places and in the right proportions. The degree to which details are rendered—some with stunning accuracy and others merely hinted at—guides the viewer's eye to the precise anatomical structures that are most important. In directing his audience's attention in this way, Brödel tells a story; he doesn't just depict a single moment frozen in time. His illustrations have a cinematic quality, the ability to focus our attention but also to show how events unfold.

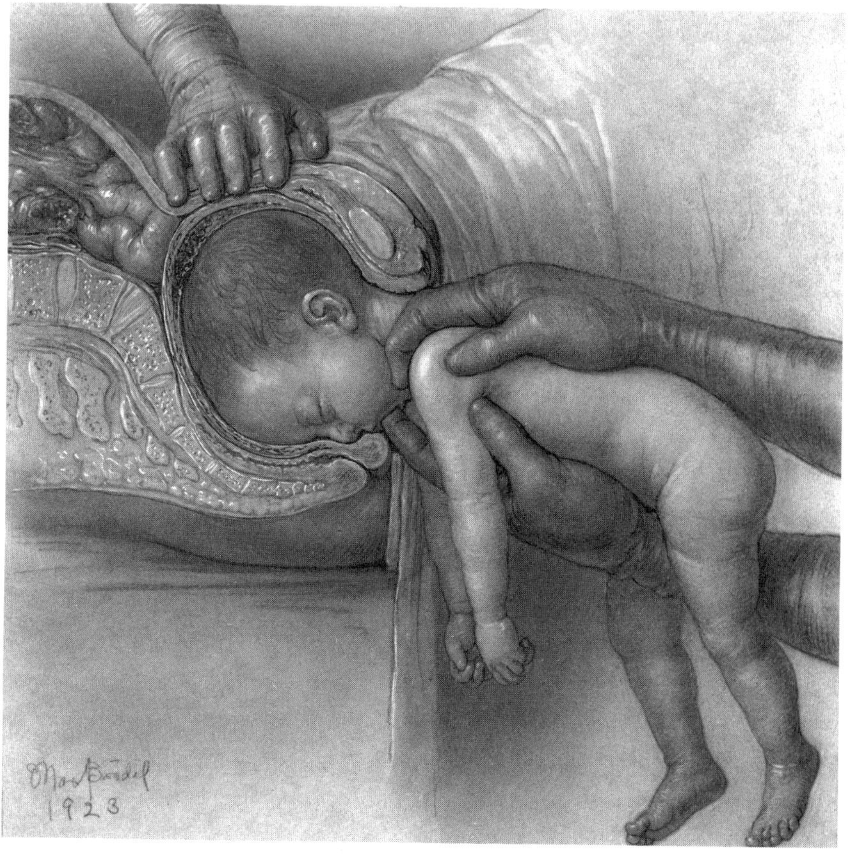

FIGURE 7.4. *Brödel's masterpiece carbon dust illustration of a breech delivery has the "wow factor" described by biographer and medical illustrator John Cody. The viewer is immediately drawn in by the detailed rendering of the baby's mouth and the obstetrician's hands (compare with the looser rendering of the mother's leg in the upper right). The cutaway of the mother's anatomy demonstrates how the baby's head fits in her pelvis. One can feel the force of the second obstetrician pressing on the mother's abdomen from above. (Wolff's carbon pencil and dust on Ross stipple board, from 1923.)*

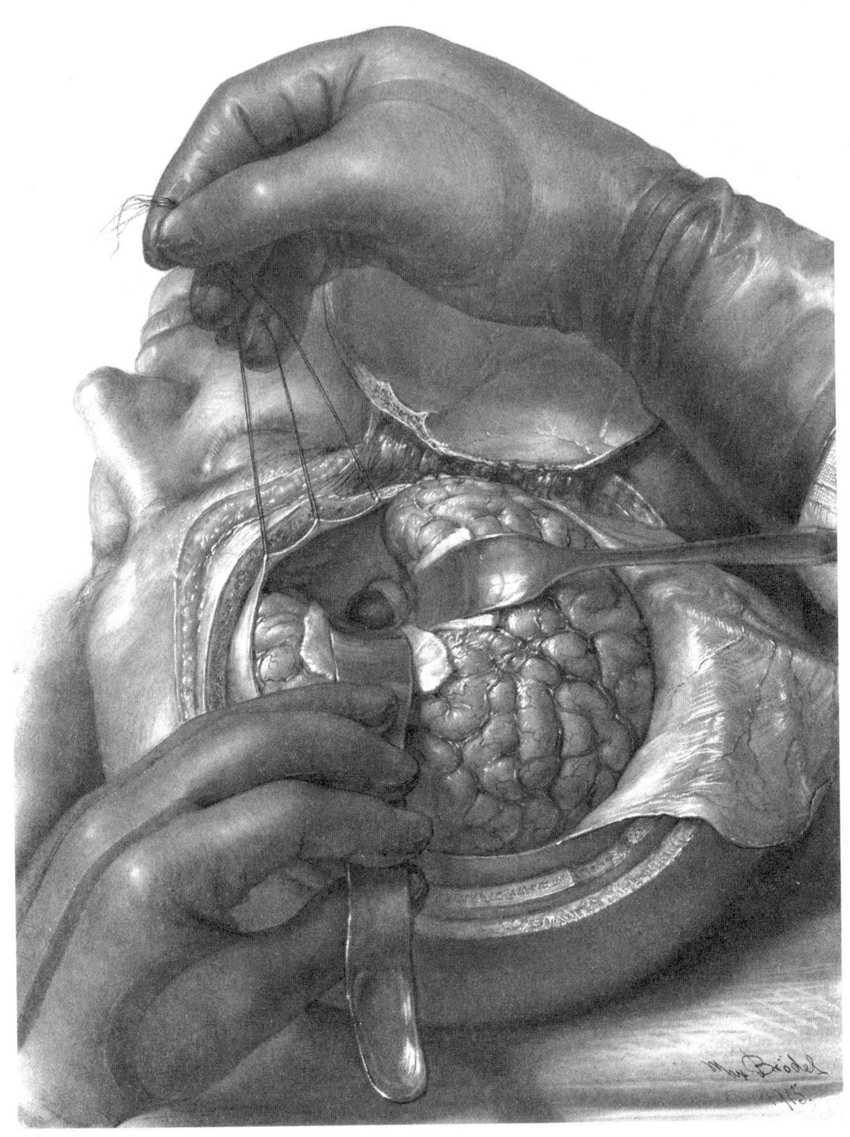

FIGURE 7.5. *The tissues glisten realistically in Brödel's carbon dust illustration of an operation on the pituitary gland. (Wolff's carbon pencil and dust on Ross stipple board.)*

Like most skilled storytellers, Brödel was concerned not only with the seen but also the unseen. An important part of his mastery was his ability to depict anatomical features that surgeons could not see with their eyes but that they needed to able to locate (or in certain cases, navigate around) to operate successfully. A prime example is his breech birth illustration (figure 7.4), where Brödel maps out the obstetrician's precise moves. He even shows the obstetrician's finger inserted in the baby's mouth, the exact method used to control the delivery of the baby's head. The techniques and concepts that Brödel illustrated were often difficult to teach in the clinic, but by making the unseen visible, he was able to convey invaluable information.

On the flipside, Brödel would sometimes "disappear" anatomy that would normally be visible in order to focus the viewer's attention on more essential elements. In doing so, his illustrations told a more informative and compelling story than even a detailed photograph could. Brödel aficionados, including several who teach medical illustration, refer to his "powerful use of . . . nothing."[28] In their own teaching, they sometimes encourage students to "think white." David Rini, a professor in the Department of Art as Applied to Medicine at Johns Hopkins, notes that "by leaving white space or downplaying detail in certain areas, the viewer's attention is focused to the surgical moment or the key point of the illustration."[29] Figure 7.6 illustrates Brödel's powerful use of white space—his selection of what to highlight and what to leave out tells an unambiguous story. The ear and chin are shown to orient the viewer, but essentially all other external anatomy is left out.

Brödel was accordingly a virtuoso at including or excluding, emphasizing or deemphasizing, anatomical details in service of his storytelling mission. He regularly employed another strategy that artists refer to as "atmospheric perspective" to accomplish his pedagogic goals. Much as the great Dutch and Flemish landscape painters of the Northern Renaissance create the illusion of haze or humidity when depicting mountains or other geographic features in the distance, Brödel would often blur objects in his backgrounds to downplay bodily structures that were not central to his instructional purpose.

We can see atmospheric perspective at play in figure 7.4, Brödel's striking illustration of a breech birth. Notice how the mother's leg depicted in the upper right quadrant quickly disappears—almost as if the limb were enveloped in a moist operating room mist. The effect is to obscure that part of the anatomy while, conversely, heightening the vividness of the baby's head, the birth canal, and even the surgeon's hands. As Cory Sandone, chair of the Department of Art as Applied to Medicine at Hopkins, has remarked, "Using conscious design decisions such as depicting unseen anatomical structures, employing white space, and exploiting atmospheric perspective, Brödel became a master of creating meaningful visual hierarchies in his illustrations to direct and focus attention."[30]

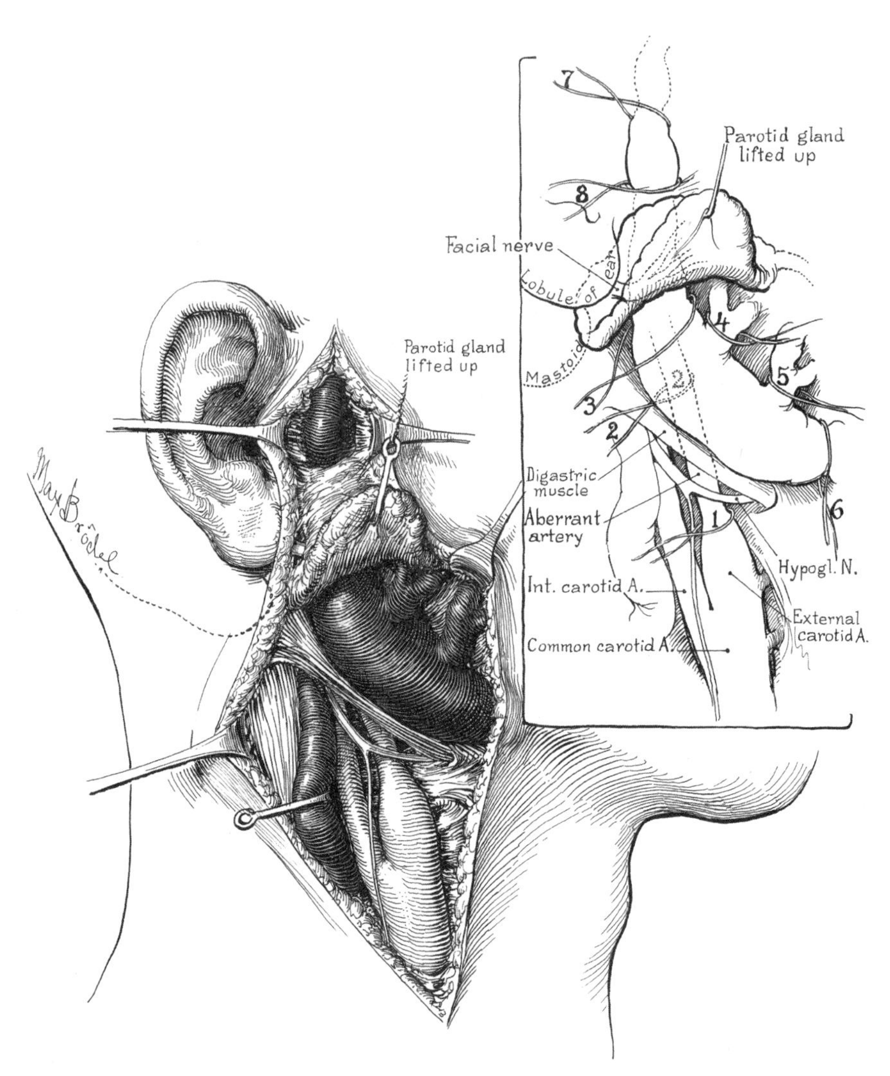

FIGURE 7.6. *Brödel's pen-and-ink illustration of surgery on an enlarged vein in the neck. Only critical details are rendered, and Brödel makes ample use of white space. (Black ink on Ross scratchboard.)*

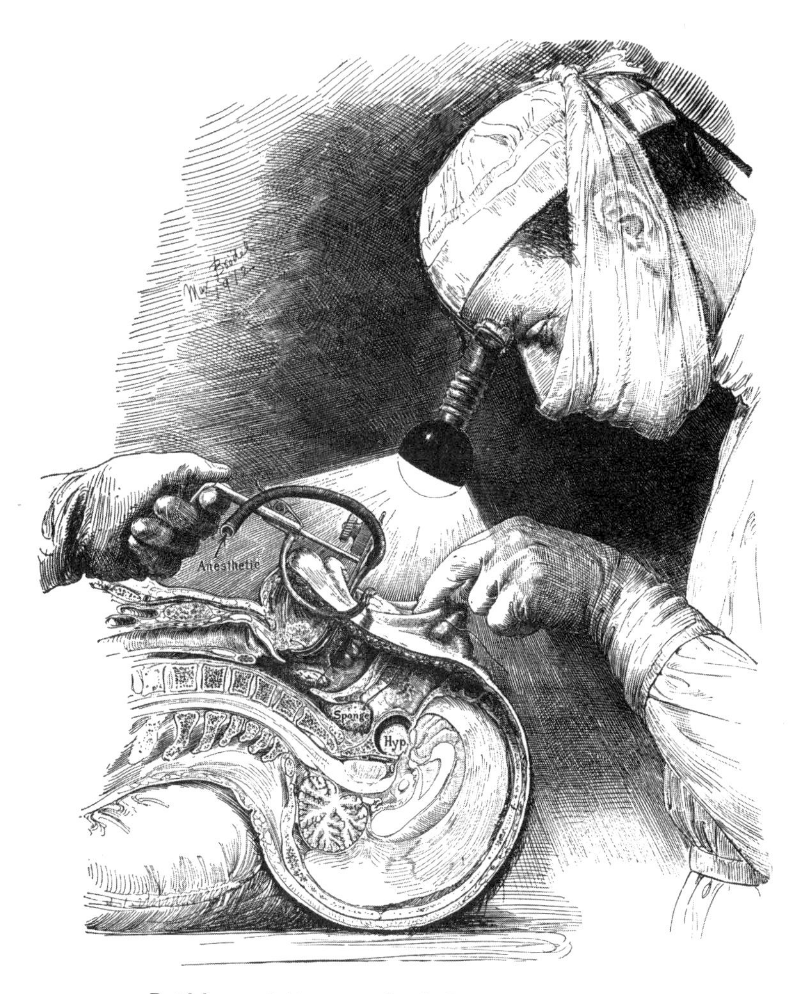

FIGURE 7.7. *Brödel created this pen-and-ink illustration of surgery on the pituitary gland for Harvey Cushing in 1912. The viewer's eye is drawn to the bold light from the surgeon's headlamp. While demonstrating the surgery, this illustration also serves as a homage to Brödel's close friend Cushing. (Black ink on Ross scratchboard.)*

From Pen and Ink to Pioneering New Techniques

Brödel was exceptionally skilled with pen and ink (figures 7.6 and 7.7). His precisely placed yet loose, casual lines convey the anatomy. As medical illustrator and professor in the Department of Art as Applied to Medicine at Johns Hopkins Timothy Phelps describes it, and as Brödel's work masterfully shows, pen and ink, "in a limited number of strokes," can convey "a sense of the artist's hand and personality—a vigor, vitality and confidence."[31] In his pen-and-ink work, Brödel uses light and dark to subtly guide the viewer's eye. Look at figure 7.6. While much of the illustration was created by applying delicate lines of ink to the board using a fine pen, the large dark vein in the center was first painted solid black, and then the paint was gracefully scratched away to highlight the white lines. This technique gives the vein full value, emphasizing it and drawing the viewer in. The use of light and dark similarly draws virtually everyone who looks at figure 7.7 to the surgeon's headlamp in the center of the illustration.

Brödel also pioneered, or at least mastered more fully than anyone previously, new mediums and new techniques. One of his greatest challenges was to find a medium that would create the highlights that characterize glistening human tissues. After multiple experiments, he found Ross Hand-Stippled Paper No. 8. This paper has a coating of bright white clay and allowed Brödel to create the illusion that the organs drawn on the page shine (figures 7.4 and 7.5). Science writer Natalie Middleton describes the steps in Brödel's "carbon dust method" as follows:

> First, Brödel would take a heavy paper covered with chalk or china clay. He would outline an image on tracing paper and then bring it into contact with the heavy paper, leaving an imprint of the overall form. Next he would take a dry brush and delicately layer carbon dust onto the outline as if he were painting. This created a rich sense of depth. As he worked, he lifted highlights out of the carbon dust with an eraser, etched in precise details using a scalpel, and darkened lines with black watercolor or carbon pencil.[32]

Brödel's three-dimensional wizardry worked. A nurse at Hopkins reportedly tried to pick up a gallstone she thought Max had placed on his favorite stipple paper.[33]

We would be misleading readers if we left the impression that every illustration Brödel created was perfectly suited to its purpose. Like many virtuosos, Brödel could occasionally get carried away. In figure 7.7, one of his illustrations of an operation on the pituitary gland, Brödel perhaps unnecessarily includes the surgeon, Harvey Cushing, as well as the operative field. In other cases, he occasionally added "painterly" details that were excessive and detracted from the real purpose of the illustration. John Cody makes this point in his incisive and entertaining "Wow! Factor" article. While acknowledging that Brödel's artwork evokes in us "admiration and an overwhelming sense of mastery and beauty," Cody argues that "a powerful Wow! Reaction is not really a plus in medical illustration. It's a handicap. Brödel had this handicap in spades."[34]

We can speculate about why Brödel occasionally embellished his illustrations beyond their basic purpose. In some instances, as in the illustration that includes Cushing, he may have wanted to reflect his close relationship with a colleague; in others it may have been purely his love of shape, form, and art itself. As Brödel himself described it, "The outlines of the various organs of the human body and their detail always possess a certain grace and interesting character all their own."[35] On balance, though, Brödel's artistic flourishes paled in comparison to his ability to capture the "surgical moment" and did not detract from his deep anatomical knowledge, technical mastery, and innovative methods he brought to his work.

A Scientific Partner

In addition to his remarkable work on Kelly's *Operative Gynecology*, Brödel, as we noted, collaborated and developed a warm friendship with the famous neurosurgeon Harvey Cushing. The partnership was a great example of the power of Brödel's artwork to create novel anatomic perspectives—perspectives previously not visualized even by the surgeon. In neurosurgery, the hard casing of the cranium and the risk that probing the brain too deeply could have fatal consequences often prevented surgeons from opening up the surgical field to a fuller view. Nonetheless, as figure 7.7 shows, with the ingenious use of a cutaway reconstruction, Brödel managed to illustrate what could not be seen, even with the surgeon's illuminating headlamp. This drawing and the rest in the series Brödel created for Cushing document how Cushing performed the delicate surgery, accessing the pituitary through the nasopharyngeal cavity. Brödel's illustrations tell a simple, clear and useful how-to story.

Even though Brödel lacked formal higher education, Cushing recognized him as a fellow scientist, writing of his neurosurgical illustrations that "few words in description of the operation are needed to supplement Mr. Max Brödel's drawings . . . As it is true of all his superb medical illustrations, they serve to make the context superfluous."[36] Cushing's view of Brödel was widely shared among his colleagues. William Halsted, perhaps the most famous of all of Hopkins's heralded surgeons in the early twentieth century, wrote in a 1918 letter to Brödel: "As I have repeatedly said to you, your drawings immortalize the lucky beneficiaries."[37] Early on in his career, Brödel also worked and formed a lifelong friendship with Thomas Cullen, who came to Hopkins as an intern on Kelly's surgical staff. Brödel would eventually illustrate several of Cullen's textbooks on surgical gynecology, and Cullen would be instrumental in establishing a medical illustration department at Hopkins, built around his immigrant friend's extraordinary skills.

Recognition and renown as a medical illustrator came at no small price. Brödel drove himself to the limits, and sometimes beyond them, to achieve his technical and artistic mastery. One example: In March 1899, in preparation for an illustration, Brödel dissected a corpse in the autopsy room with ungloved hands. Later, his left hand began to swell and turn red. The deceased patient had died of a streptococcal infection.[38] As the infection spread up his arm, Brödel was repeatedly taken to the operating room to drain pus and alleviate the swelling. This was in an era before antibiotics. By Brödel's own account, "my life hung by a thread."[39] As he faced this grave threat, Brödel created a series of twenty sketches documenting the pain and progressive loss of sensation in his arm (figure 7.8). He eventually recovered from the infection, but his arm was still partially paralyzed. (The infection was in Brödel's nondominant left arm, which was nonetheless an indispensable support for just about every task he performed with his right.)

Desperate to regain the use of his arm, Brödel conducted a series of detailed anatomic studies on cadavers and determined the exact point where his ulnar nerve was pinched by scar tissue. Halsted surgically freed the nerve in just the way Brödel's research and drawings suggested.[40] Brödel regained the use of his arm, and the trajectory of his career, which had been interrupted for about six months, resumed.

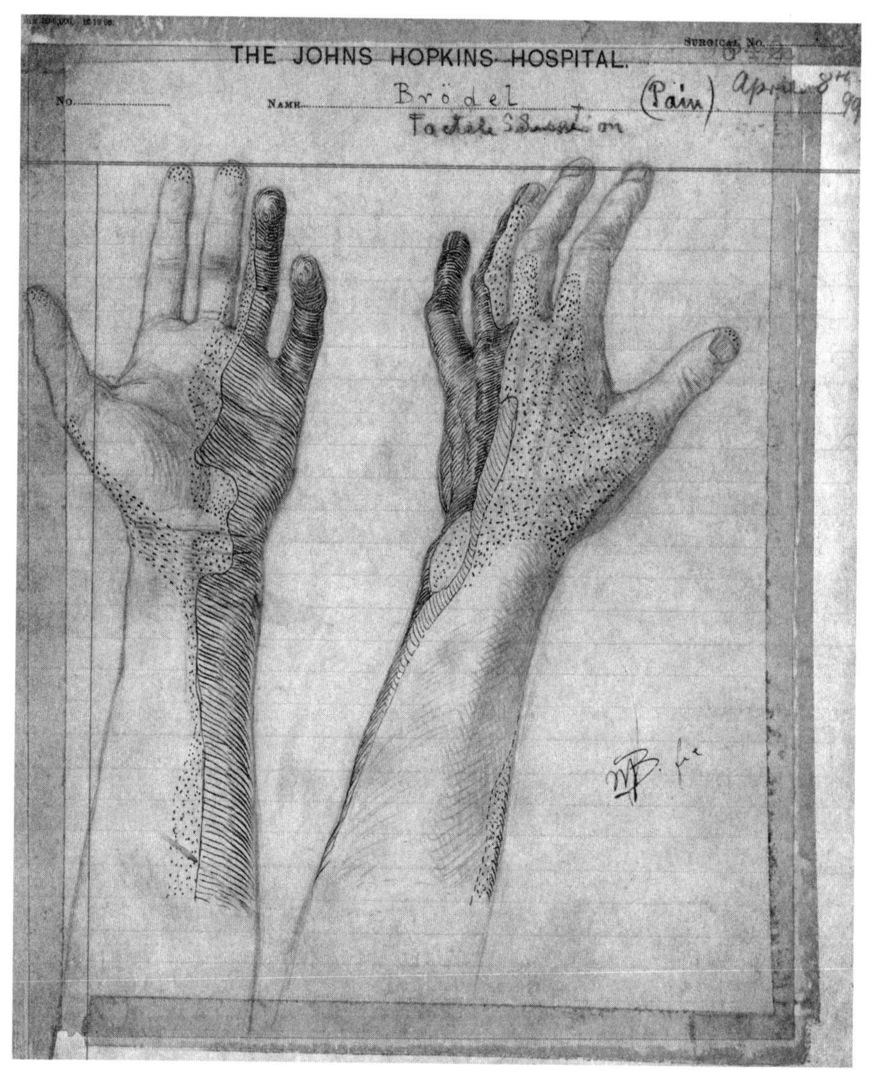

FIGURE 7.8. *Brödel documented the progressive loss of sensation as a life-threatening streptococcal infection spread up his arm.*

Professionalizing Medical Illustration: The Department Brödel Built

Despite his growing reputation and influence, for the first ten years of his career at Hopkins, Brödel was not an employee of the university. Instead, his salary was paid by Howard Kelly, the renowned gynecologist for whom he produced illustrations. In 1911, Brödel's colleague and close friend Thomas Cullen went to Henry Walters, the Baltimore businessman and philanthropist whose bequest founded the city's Walters Art Museum, asking him to support Brödel's work. Walters responded to Cullen's request by writing, "I am not interested in medical illustration . . . I took twenty lectures in medicine at Harvard and nearly vomited my boots up." Nonetheless, Walters agreed to fund Brödel, noting in his letter to Cullen: "However, I appreciate the value of medical illustrations and will give the Department of Art as Applied to Medicine 5000 dollars a year for three years."[41]

The department, the world's first formal training program for medical illustrators, was established in 1911. Brödel became its first director, eventually rising to the rank of associate professor in the School of Medicine. In 1921, Walters endowed the department with an additional gift of $110,000. Brödel made his profound contributions to art and science at Johns Hopkins despite only having the equivalent of a high school degree.

Behind these facts lies an intriguing backstory. The story highlights Brödel's reputation as the best medical artist of his or any generation. It also points to repeated efforts by rival medical institutions to poach top talent from Hopkins early in the twentieth century. In 1904, Brödel was invited to move to the Mayo Clinic. Physician-in-Chief William Osler apparently helped dissuade him from leaving. The following year the university's board voted to appoint Brödel an instructor *without pay*. Charles Mayo visited Baltimore in 1906 and again tried to recruit Brödel. His brother William Mayo wrote Brödel: "We will never be happy without Mr. Brödel."[42] Kelly immediately upped his salary to $5,000, and Brödel agreed to stay. In 1910, the Mayo brothers tried yet again, an effort apparently thwarted once and for all by Walter's gift the following year that established the Department of Art as Applied to Medicine.

Between 1911 and Brödel's retirement in 1940, he trained over two hundred artists to become medical illustrators (figure 7.9). He was a one-man faculty, teaching courses in anatomy, physiology, and pathology as well as various illustration techniques, including pen and ink, watercolor, and his signature carbon dust process. Brödel also continued to complete assignments for his medical colleagues. His preferred method of teaching was hands-on, relying primarily on individual tutoring sessions and small group demonstrations. Brödel used no written syllabi, nor formal curricula, though one former student claimed the logical progression of topics covered was apparent when she inspected the portfolios of more advanced students.[43] Typically, students studied in the department for two years, but there was no fixed program length or graduation requirement. Brödel would advise students when he felt they were ready for employment and frequently provided them with a letter of recommendation. No diplomas were awarded, and by all accounts graduation was largely a matter of "a handshake."[44]

FIGURE 7.9. *Brödel teaching (standing, left) in the classroom, circa 1917.*

As Brödel described it, the program's purpose was "to train new generations of artists to illustrate the medical journals and books of the future and to spare them the years of trials and disappointments of their self-taught predecessors."[45] True to his talent at visualization and his exceptional artistic skills, Brödel taught his students to "study the subject in the form of pictures, not of words."[46] Brödel's style as a teacher was informal but highly engaging. Perhaps the best example is one of the remarkable techniques he employed in his anatomy courses. He had two skeletons and placed each behind a large pane of ground glass. Beside each pane he posed a nude male model—one heavy and the other thinner. He would then draw on the glass, first adding muscles to the skeletons, and then progressively building up with his wax pencil each layer of tissue until there on the glass were beautiful illustrations of each of the two seated nudes.[47] Ranice Crosby, Brödel's biographer and also his former student and eventual successor as director of the Department of Art as Applied to Medicine at Hopkins, wrote that Brödel's trainees were captivated by "his merry, spontaneous personality, his lack of arrogance and pomposity, his inventiveness."[48]

The students Brödel trained helped establish medical illustration as an academic discipline and an indispensable component of scientific medicine. In the 1940s and 1950s, seven of his students, according to Crosby and Cody, established medical illustration programs at other institutions, including Massachusetts General Hospital, the University of Rochester, the University of California at San Francisco, and the University of Toronto. (Interestingly, and perhaps a sign of Brödel's stature and influence, none of these competing programs were set up until after his death.) Four more of Brödel's students went on to lead established medical illustration programs, including two of his former pupils who were, in turn, occupants of Brödel's post at Hopkins.

The Private Brödel: Close Friendships but Ever the Immigrant Outsider

As this account of Brödel's teaching style suggests, there was something enchanting about his personality. Though Brödel thwarted his father's wish that he pursue a career as a concert pianist, he never lost his love for the piano and playing. Perhaps it was this affinity for the unspoken language of music that made him such an engaging and empathetic companion. Many of Brödel's closest relationships had a musical connection.

Music even played a role in his choice of a life partner. In 1902, Brödel married Ruth Harrington, who, after completing graduate work in anatomy at Smith College, had come to Hopkins as a medical illustrator. The couple had been introduced by Howard Kelly, Brödel's employer and mentor, who probably recognized the young people's shared interest not only in art but also in music. Something of a matchmaker, Kelly had invited Brödel and Harrington to spend part of a summer at his retreat on Lake Ahmic in Ontario. Leaving nothing to chance, Kelly arranged to have a piano delivered to the rural camp. Brödel and Harrington spent several romantic interludes playing four-handed arrangements. In her memoir, Ruth Brödel wrote: "One [Schumann] composition afterward remained in [our] memories, indelibly associated with that peaceful sylvan scene, the buzzing of bees, and sunshine warming up the fragrant morning air."[49]

Kelly's scheme worked. Ruth and Max announced their engagement to their host and his summer guests a few days later and were married the following December. The Brödels had four children, one of whom tragically died of scarlet fever as a toddler. Their oldest daughter, Elizabeth, went on to become a medical illustrator, a brave choice with such a famous father. Carl, the Brödels' son, earned a PhD from Hopkins and pursued a career as a geologist. Their fourth child, Elsa, became an art teacher.

We have seen how Thomas Cullen, the gynecologist and future chief of gynecologic pathology at Hopkins, was instrumental in securing the funding to establish the Department of Art as Applied to Medicine. Cullen and Brödel were already close personally as well as professionally. In fact, they enjoyed a lifelong friendship. An indispensable part of their bond was language. Early on, in an effort to become fluent in each other's native tongue, Brödel spoke only in English to Cullen, and Cullen spoke exclusively in German to Brödel. Their pact lasted until Brödel died forty-seven years later. Cullen wrote: "As he spoke English and I German we very frequently had temporary misunderstandings, but as our knowledge of the two languages became greater we got along perfectly."[50]

In addition to music, Brödel loved nature and had a special gift for portraying it. Brödel eventually bought a cabin on Lake Ahmic in Canada. For Brödel, there was a similarity between the work of the medical illustrator and the observer of nature, though the latter had a far more relaxing and contemplative job. In one of his department's annual reports, he even wrote that an interest in tiny creatures that creep and fly was a good predictor of a student's aptitude for medical illustration.[51] Predictably, Brödel had exceptional ability when it came to visualizing nature. His friend the journalist and cultural critic H. L. Mencken put it this way: "He did not see only the woods, he saw through the woods, as it were, into their mysterious and limitless depths."[52]

One of Brödel's special talents was collecting bracket fungi (large mushrooms) that grew on tree trunks around Lake Ahmic and using a fine pinpoint to etch them with beautiful illustrations of the forest surrounding his cabin. His friend Cullen described the process this way:

> When we took long trips through our woods, Max would be on the lookout for fungi. These were often a foot or a foot and a half long and were attached to the sides of the trees. They were usually brown on top, pure white on their under sides. Brödel would take these fungi home, carefully protecting the white under sides of the fungi. On reaching home he would, with a pen, a pin or some other sharp instrument, etch a charming scene on each fungus . . . I have never seen their equal.[53]

Ruth Brödel noted that Max would cleverly adapt the scene he was etching to take advantage of the particular contours of the fungi, using bumps in the surface to more realistically depict trees, boulders, or other features of the forest. To this day, the Max Brödel Archives in the Department of Art as Applied to Medicine at Johns Hopkins still houses some fine examples of this amazing artwork (color plate 11).

In a more raucous vein, his journalist friend Mencken invited Brödel to join the Saturday Night Club, a varied group of artists, musicians, intellectuals, and freethinkers who gathered every week in Baltimore for two hours of playing music followed by two hours of drinking and smoking cigars (figure 7.10). It was Mencken, notorious for his sardonic wit, who was reported to have said, "No one in the world, so far as I know—and I have searched the records for years, and employed agents to help me—has ever lost money by underestimating the intelligence of the great masses of the plain people."[54] Mencken and Brödel played the piano together at these weekly gatherings, with the journalist noting that Brödel played with tremendous joy, wearing a smiling, charmed expression. True to form, Mencken added that Max was "a pyano-thumper" who had played "more than 900 billion notes in the past fifty years, no two precisely alike."[55]

FIGURE 7.10. *The Saturday Night Club. Brödel (second from right in white shirt) would play piano with H. L. Mencken (far right). Beer, food, and lively discussions followed.*

Prohibition went into effect on January 17, 1920. Mencken responded in word by railing against it in his newspaper column and in deed by brewing his own beer. He quickly enlisted Brödel. Mencken was able to obtain yeast from the Löwenbräu Brewery in Germany. He soon discovered that Brödel had connections in the medical school who knew a lot about microbiology, and Brödel quickly became the best brewer in Mencken's circle of bootlegging beer makers. Noting that Brödel's beers had "a genuinely professional smack,"[56] Mencken approvingly wrote, "in this humanitarian work [he] had the aid of various Johns Hopkins colleagues."[57] Brödel kept detailed notes on his beer-making recipes and technique, some of which have been preserved. Judging by the quantities of ingredients listed (for example, "3 lbs. of corn sugar"), Brödel was not bashful about brewing up a large batch of bottles.[58]

Brödel was wonderfully modest. Cullen, perhaps his closest friend, described him as "a kindly, curly-haired man of quiet demeanor."[59] In an article commemorating the one-hundredth anniversary of the founding of the Department of Art as Applied to Medicine, John Cody wrote: "Publicly, Max Brödel downplayed the beauty of his art, seeming to take that casually for granted."[60] Brödel's view of himself was equally self-effacing. Responding to an invitation for a dinner in his honor in 1938, Brödel told his hosts "that you make your celebration an occasion to turn the limelight on my humble person fills me with panic."[61] At the same dinner, Brödel called medical illustration "the most interesting, thrilling and satisfactory work imaginable."[62]

Despite his close friendships and the accolades bestowed on him by professional colleagues, Brödel remained in many respects a man without a country. In 1898, four years after arriving in Baltimore, he returned to Germany, only to find himself a stranger there. Physically, the chronic bronchitis that had afflicted him in his youth flared up again when he reached his hometown of Leipzig, only this time more severely. He recognized that the return of the illness likely meant separating himself from his family permanently. Anxious at that prospect, he wrote about his longing to return to Baltimore. Brödel returned to Leipzig for another visit in 1911, but still experienced the same low spirits. By this time, he had been a US citizen for more than five years. He wrote his wife, Ruth, about his native Germany, "The land and its people have changed, or else I have changed, and we don't agree exactly."[63]

Although he was an American citizen and had many close personal and professional ties in Baltimore, Brödel never felt embraced by his adopted country. Not only by his own account, but in the view of others, including Mencken, Brödel had "a very unpleasant time of it in World War I, and would have suffered almost as much in World War II if he had lived."[64] Even his easy, affectionate relationship with Harvey Cushing was temporarily soured by Brödel's perceived sympathies for Germany. Brödel wrote Cushing in 1914: "Harvey dear, don't believe the *All-lies* [italics ours] reports, remember that one of your most devoted friends belongs to the other side. Give the Germans half a chance."[65] Cushing replied with what Brödel would call a "beastly" letter. Brödel, clearly stung, dashed off another letter telling Cushing, "I do not deserve to be disowned by those I love."[66]

Brödel is known to have expressed both anti-Black and anti-Semitic sentiments. Brödel wrote disparagingly to his brother-in-law Hermann Schäfer of "the 'black element,' the g-d- negroes" and "the conscienceless, felonious blacks who are alike as one egg is to another."[67] In this instance, he was describing the role of Black people, apparently recruited and paid by unscrupulous politicians, in the corruption and intimidation that routinely occurred on Election Day in Baltimore. Whatever the context, Brödel's racist remarks are shocking. He also harbored a milder animosity toward Jews. Again writing to Schäfer, this time in connection with making his relative a loan, Brödel followed up the news that he would charge him no interest with this snide trope about Jewish business practices: "The daring and the cunning that our Semitic fellow men apply in the battle for existence cannot be expected from you."[68]

While not shying away from describing Brödel's racial, religious, and ethnic prejudices, biographers Crosby and Cody work hard to set them in context. They point to the culture in which Brödel grew up in his native Leipzig. They note that Bismarck's recent unification of Germany had made Jews and other groups who spoke a different language and celebrated their own traditions objects of suspicion and often caused these groups to be viewed as an outright threat to the nation's destiny. Crosby and Cody also take pains to situate Brödel's bias against minorities in the "climate of feeling that surrounded him" in early twentieth-century Baltimore.[69] Just below the Mason-Dixon Line, the city was distinctly southern in its attitudes toward race and religion. This effort to place Brödel's views in context does not excuse him. It is also a paradox that Brödel's own painful experiences with being an outsider in his adopted country and the target of anti-foreign prejudice appeared to have little or no impact on his empathy for others facing similar challenges.

There is every indication that Brödel was a loving and supportive husband and father to his wife, Ruth, and their children. Yet, here too, Brödel's views about women and their roles in the larger society were distinctly Victorian. Discussing female medical illustration students, Brödel opined, "women get married and abandon the work, while matrimony in a man increases his determination to succeed." He added, "The medical profession has always shown preference for male illustrators."[70] Not surprisingly, Brödel opposed women's suffrage. As discussed in our introduction, the biases held by the ten men and women portrayed in this book provide all of us with an opportunity for us to reflect on our own ingrained biases—conscious and, perhaps more to the point, unconscious.

The bronchitis that affected Brödel in his youth and seemed to recur every time he visited his family in Leipzig was only one of a series of serious illnesses that the normally vigorous Brödel endured throughout his life. We mentioned that as a child, Brödel suffered a severe case of smallpox and that his nearly fatal arm infection occurred five years into his career at Hopkins. In 1904, Brödel developed an acute case of typhoid. He was hospitalized for eight weeks, treated with alcohol rubs to lower his body temperature, and again almost died. Almost a decade and a half later in 1918, Brödel developed a mass adjacent to a tooth. Halsted diagnosed it as cancer. The mass was resected and fortunately found to be benign. Some days later, presumably after Brödel had received the news that that growth was not malignant, his friend Mencken and some fellow members of the Saturday Night Club visited Brödel in the hospital. They told the staff that they were there to conduct "a coroner's inquest."[71] Pulling long faces, they offered the mildly anticlerical Brödel a Gideon Bible.

Despite being so meticulous in his art, Brödel apparently could be careless in his work routines. In 1921, he was using a solution of potassium cyanide to bleach a sketch. Unnoticed by Brödel, a steady wind was blowing in from the window. Had it not been for a student who happened to enter the room and found Brödel barely conscious and collapsed at his drawing board, things might have ended in tragedy.

A Modern-Day Leonardo

Max Brödel died of pancreatic cancer on October 25, 1941. He was seventy-one. There were fifty-six honorary pallbearers at his funeral. The majority were physicians, a testimony to the rich trove of extraordinary illustrations that had aided and disseminated their medical accomplishments. It was a reflection as well of their deep gratitude to the humble but exceptionally talented artist who had created them. In an appraisal of Brödel's career filled with insight unusual even for a talented journalist, Mencken wrote in the *Baltimore Sun*:

> There had been, of course, many anatomical artists before him, and some of them were draftsmen of the first caliber. But it remained for him to make medical illustrating an authentic and indispensable branch of the new medicine. He saw in every piece of work confronting him, not as a problem in mere drawing, but a problem in research and elucidation . . . In even the least of his long line of superb drawings there was visible an anatomic knowledge that was both precise and enormous; in the best there was a quality of scientific imagination that made them contributions of the first importance to the study of the human body in health and disease.[72]

It was Ranice Crosby, whose superb biography of Brödel with John Cody helped us shape this chapter, who called him "the greatest anatomical artist since Leonardo da Vinci" and "a creative genius, the likes of which would never come again."[73] Crosby was in a position to know since she had studied under Brödel.

Longtime colleague and close friend Thomas Cullen said of Brödel, "He was a born artist and during his forty-eight years in Baltimore revolutionized medical illustrating in the United States and Canada; his work has reached even the uttermost parts of the earth. No other man who has ever lived has done as much to improve the beauty and accuracy of medical illustration . . . Brödel will be recognized as the greatest medical illustrator who has ever lived."[74]

When we take the long view of Brödel's life and achievements, it is worth considering these facts about some other extraordinary contributors to science. Among the fifty-nine Nobel prizes awarded to US citizens in medicine or physiology, physics, and chemistry between 2000 and 2019, 40 percent were won by immigrants to this country.[75] Max Brödel's contributions to scientific medicine exemplify the enormous impact immigrants have had on Hopkins and on American medicine. Brödel's life and career also underscore, as we have seen in the case of Mary Elizabeth Garrett and will learn in our final chapter on Vivien Thomas, the huge impact that individuals who were not trained as physicians have had in the reinvention of American medicine.

8

DOROTHY REED MENDENHALL

Hardship and Discrimination

The National Cancer Institute's description of Hodgkin lymphoma suggests the magnitude of the twenty-eight-year-old Dorothy Reed's accomplishment. NCI's website states that the disease "is marked by the presence of a type of cell called the Reed-Sternberg cell."

Reed was one of fourteen women in a cohort of forty-three students admitted to the Johns Hopkins School of Medicine's fourth entering class in 1896. Her presence was a direct result of Mary Elizabeth Garrett's unyielding approach to philanthropy that compelled the school to admit women on an equal footing with men. (Harvard Medical School didn't admit women until 1945.) In 1901, having just completed a prestigious internship in medicine under William Osler, Reed accepted a fellowship in William Welch's pathology laboratory. She set to work on Hodgkin lymphoma, also known as Hodgkin disease. A cancer of the white blood cells known as lymphocytes, Hodgkin disease typically manifests symptoms including swollen lymph nodes (glands), fever, and night sweats. Although it is now highly treatable, in Reed's day, Hodgkin disease was nearly always fatal.

A few years before Reed started her work, the Austrian pathologist Carl Sternberg had minced lymph nodes from patients with Hodgkin disease and inoculated the fragments into guinea pigs. He reported that the animals developed tuberculosis. As a result, many pathologists at the time believed that tuberculosis was "masquerading as Hodgkin's disease."[1] Never one to shy away from a debate, even with an established authority, Reed carried out her own research with rigor, energy, and creativity. She studied the tissues of patients with Hodgkin disease under the microscope, a relatively new research tool, recorded the results of tuberculosis skin tests, and even repeated Sternberg's experiments inoculating animals with ground-up lymph nodes to see if they developed tuberculosis.

In Reed's experiments, guinea pigs injected with a solution containing minute pieces of lymph nodes from patients with Hodgkin disease failed to develop tuberculosis. Her microscopic studies highlighted marked differences between the two diseases. She wrote of Sternberg's idea that she could not "understand his conception . . . of the *tubercle bacillus,* which he suggests is the cause of such a growth . . . the two processes are easily distinguished."[2] In unambiguous terms, she concluded that tuberculosis had no direct role in Hodgkin disease. She also made the important observation that patients with Hodgkin disease had a weakened immune system.

Reed's contributions to the medical science surrounding Hodgkin disease went beyond disproving Sternberg's hypothesis that Hodgkin disease was caused by the tubercle bacterium. Through careful microscopic studies, Reed observed a characteristic giant cell, frequently containing multiple nuclei, in the lymph nodes of patients with Hodgkin disease, that others, including Stenberg, had seen but not emphasized. Her report noted, "Large giant cells, containing one or more large nuclei were numerous."[3] The report was more than sixty pages long (including a detailed appendix that contained clinical data, patient photos, pathological findings, and a discussion of laboratory methods), written in an authoritative style, and skillfully illustrated in Reed's own hand. Based on this groundbreaking research, and with the attention it later received in a former colleague's pathology textbook, she ultimately won international recognition. The telltale diagnostic marker of Hodgkin disease is now named the Reed-Sternberg cell.

What was Reed's reward for her pathbreaking scientific accomplishment? By all rights, Reed's important discovery should have put her on track for a promotion, but the year was 1902, and Reed was a woman. As her yearlong fellowship was ending, she asked Welch about her prospects for obtaining a faculty appointment. "I explained," she wrote, "that the man who had had the fellowship just before me had done no research but had been made an assistant [professor] in pathology the next year . . . Why not I?"[4]

Although he was a supporter of coeducation and welcomed several women into his lab, Welch answered Reed by telling her that there was no precedent for having a woman faculty member in the medical school and that he expected stiff opposition to such an appointment. His words were a body blow to Reed. She wrote in her memoir, "Suddenly, as I saw what I had to face in acceptance of injustice and in being overlooked—I knew that I couldn't take it. And I told Dr. Welch that, if I couldn't look forward to a definite teaching position even after several years of apprenticeship, I couldn't stay."[5] Reed was not alone in being defeated by the glass ceiling—more like a concrete roof—that prevented women from rising in the medical school. In the same era, her colleague and medical school classmate Florence Sabin, who would go on to a distinguished career at the Rockefeller Institute and become the first woman inducted into the National Academy of Sciences, also had her career at Hopkins thwarted. Although Sabin became the first woman appointed to the faculty of the School of Medicine in 1902, she was passed over in favor of one of her former male students for the position of department chair when her mentor died. In all, Sabin waited fifteen years before being promoted to full professor.

Pioneering in an Age of Transition

Sexism complicated Reed's life at almost every turn. There was her family's expectation that Dorothy deploy her beauty, polished manners, and privileged upbringing to find a suitable husband. There was also society's preference that she forgo a higher education and a career in favor of marriage and motherhood. And finally, there was the medical community's contradictory and damaging attitudes—demeaning, contemptuous, and leering in various combinations—toward women physicians.

Despite these obstacles, Reed came to adulthood at the right time. In the last years of the nineteenth century, a rapidly expanding and increasingly automated industrial economy, the rise of public education as well as women's colleges, and, above all, the strengthening of the suffragist movement combined to challenge established orthodoxies and create new opportunities for women. Reed used every ounce of her considerable talent—and toughness—to seize these openings, often paying a steep emotional price.

In her memoir, Reed claimed that, looking back, family burdens and financial pressures in her early years "forged my character into iron. Any sweetness I may have had—turned to strength."[6] Reed's emotional self-portrait may have been an exaggeration. She was known for her tenderness with children. And she devoted much of her career to caring for those who could least afford it. Nonetheless, at almost every stage of her life, Reed struggled mightily to overcome obstacles not of her own making. In this chapter, we explore Reed's early years, her struggles and achievements at Johns Hopkins, her marriage and its challenges of balancing family and career, and her legacy as a pioneering physician-scientist and advocate for women and children.

Reed's story is worth telling for other reasons. Like Helen Taussig and Vivien Thomas, whose lives we recount in our final two chapters, Reed's history at Hopkins reveals the sexism and racism that marked the institution and virtually all medical schools at the time. In the current moment, when structural discrimination is very much a part of our national consciousness, we can learn vital lessons from their stories of personal and professional sacrifice.

An Early Streak of Independence

Dorothy Reed was born in Columbus, Ohio, on September 22, 1874. Her earliest years were spent growing up in a well-to-do and locally prominent family. Reed's hardworking and ambitious father, William Reed, had struck out for the Midwest as a young man from his home in Massachusetts, eventually securing work in a shoe store in Columbus operated by Hannibal Kimball. Will Reed's enterprising character impressed his employer, and he was soon a partner in the business. A few years later, in 1865, Reed married Kimball's daughter Adaline Grace (always known as "Grace"). Not content with his improving fortunes, Reed formed a new partnership to start a shoe and boot manufacturing business. The venture prospered and eventually employed 120 people at its headquarters in Columbus and held assets valued at $600,000 ($13,842,000 in today's dollars).[7] By the time Dorothy arrived, the couple lived in a fine home on Town Street in Columbus. Dorothy was the youngest of the Reeds's three children, trailing her brother William by five years and her sister Elizabeth by three. She cherished her father. Years after his death, she called him "[my] ideal almost hidden companion, and my touchstone for conduct and important decisions."[8]

Dorothy's world was turned upside down at the age of six. She awoke one night to discover that her father, who had been chronically ill with diabetes and tuberculosis, had died in his sleep. Grace Reed, suddenly the sole head of the household, had been a loving wife and fashionable hostess, bringing style and social connections to her husband's burgeoning business. She was, however, completely unprepared to manage her newfound personal wealth, though she surely enjoyed the financial freedom it gave her. As an adult, according to her biographer Peter Dawson, Dorothy described her mother as "'a woman of refinement but no culture,' of great generosity but superficial values, and 'the kindest woman I have ever known [but with] the poorest judgment.'"[9] Dorothy was still a young child, but, as Dawson points out, the stage was being set for a lifelong struggle—and eventual role reversal—between Grace Reed and her youngest daughter.[10]

Like most Victorian mothers with social standing, Grace Reed's fondest hope for her daughters was that they would grow up to possess the charm and refinement to marry a wealthy husband. Grace's oldest daughter, Elizabeth, with her striking beauty, elegant manners, and musical gifts, was on her way to fulfilling her mother's ambition. Dorothy was another matter. From an early age, and especially after her father's premature death, she resisted Grace's efforts to make her a proper young lady. She preferred, in her own words, roaming outdoors "like an unbroken colt."[11] Even within the confines of her home and family, Dorothy preferred the company of imaginary friends.

After both girls recovered from severe cases of tuberculosis, probably contracted years earlier from their father, Grace took Dorothy on a vacation to Europe. (Elizabeth returned to boarding school.) With few routines and almost all her time spent in the company of her cousins, who were young adults, Dorothy became quite willful and argumentative. On the French Riviera, a moonlighting gendarme had to be hired to watch over Dorothy while the others took a break from her antics. She spent her time trying to elude him.[12] At wit's end, Grace Reed vowed to hire a governess when she returned to Columbus.

Miss Anna C. Gunning proved to be Dorothy's salvation.[13] Gunning, "an Irish woman of real culture and educational values," was a plain-faced retired high school teacher with a keen mind and a formidable determination to educate her pupil.[14] At first, Dorothy strenuously resisted Gunning's efforts to introduce her to subjects like grammar and mathematics. (Of the latter, Dorothy declared, "Why do I have to learn this stuff—what does it have to do with life?")[15] Over time, however, Dorothy's native intelligence and drive to succeed when she was challenged took hold. Dorothy's command of subjects ranging from geometry to French steadily improved. Miss Gunning wisely chose to let Dorothy select the literature they read and managed to convert her into a voracious reader.

Anna Gunning worked with Reed for three years, but the two stayed in touch far longer. Much later, Reed wrote: "Thinking over this period in my life, I cannot be grateful enough to this quiet, austere, unattractive woman who came into my life at the crucial moment . . . She literally set my feet solidly on the paths of higher learning and inspired me with her real love of knowledge and respect for mental attainment."[16]

Near the end of Gunning's tenure, Dorothy traveled with her to Berlin, where Dorothy's mother and older sister were already staying on a trip designed to improve Elizabeth's considerable skills as a pianist. There, Dorothy and Gunning met the editor of the *Cincinnati Enquirer* and through him the daughters of the president of Smith College. Grace Reed's vision of a higher education for Dorothy began and ended with an extended stay in Paris, where her younger daughter might attend embroidery school and improve her French.[17] Dorothy detested the idea. Instead, using their newfound contacts, Gunning arranged for Dorothy to take the entrance exams for Smith. To Dorothy's astonishment, except for geography, she passed all her exams, the first she had ever taken. In the end, the geography requirement was waived, and Dorothy was admitted to Smith in the fall of 1891.

Smith College was founded in 1875 by its namesake, the philanthropist Sophia Smith, to provide women with the opportunity to obtain educations "equal to those afforded young men in their colleges."[18] Like Bryn Mawr College, established ten years later and eventually led by Mary Elizabeth Garrett's close friend M. Carey Thomas, Smith was in the vanguard of the women's education movement. Creating colleges for women was controversial in a male-dominated society that feared higher education would deprive women of their feminine qualities and their traditional childbearing and childrearing roles. Smith offered a surprisingly broad curriculum, including physics, chemistry, biology, and astronomy. The college also provided women with an unprecedented degree of freedom. Victorian women intrepid enough to desire a college education were no shrinking violets, and Reed and her Smith classmates took full advantage of the opportunities.

Reed initially enrolled in Smith's literary studies program, which encompassed Latin, English, French, German, mathematics, and history, along with the mandatory Bible course taught by Smith's president, an ordained minister. In her sophomore year, Reed became disenchanted with her English and French classes and decided to enroll in a course in general biology. The subject and the young professor, Harris Wilder, dazzled Reed. "This is it," she recorded in her memoir, "this makes sense—this is what

I have been waiting for all these years."[19] Wilder became the third of the triumvirate of the most important people in her life along with her father and Miss Gunning.

Reed's later years at Smith were filled with academic and extracurricular accomplishments, but also burdened with family and financial pressures. In her junior year, Reed was chosen as one of the two most attractive women in her class and was also selected to serve as an usher at commencement. It was perhaps the first time Reed, long overshadowed by her more glamorous older sister, realized she was pretty—and saw the complications that go along with beauty. Reed wrote, "I never could see why you should take any credit to yourself for having them [good looks]."[20] A more serious challenge as Reed neared the end of college was her mother's spendthrift habits. Returning to her family's summer home in upstate New York at the end of her junior year, Reed found the executor of her father's estate staying there. Grace Reed's extravagance and her failure to rein in her son Will's even more profligate ways was bankrupting the family.[21] Someone practical and hardheaded needed to take charge. That someone was clearly Dorothy. A series of anguished and bitter arguments ensued, but in the end Grace and Elizabeth surrendered control over their shares in the family fortune to Dorothy. At nineteen, she knew little if anything about managing an estate, but with the executor's guidance, she quickly learned. For the remainder of her life, Dorothy astutely managed her family's wealth.

By attending college, Reed had for the moment escaped her mother's ambition to make her youngest daughter a desirable "catch" for an eligible and wealthy bachelor. But Reed's early assumption of financial responsibility for her entire family made one more thing clear: She needed to make a living, and a good living at that. Since biology was the subject that inspired her most at Smith, Reed was more than intrigued when a Smith acquaintance introduced Reed to her younger brother, a Harvard man visiting Northampton for the day. The young man described, in Reed's words:

> A remarkable school of medicine being started in Baltimore where every student accepted would be a graduate of an accredited college with special preparation in science and a reading knowledge of French and German. My friend's brother had

> been accepted and was to enter in the fall. My interest was aroused enough to ask if women were to be admitted and how information concerning Johns Hopkins Medical School could be obtained.[22]

It was a measure of Reed's decisiveness and her ability to size up her opportunities that she wrote that same night to William Welch, dean of the medical school, inquiring about admission.

When Welch replied, he informed Reed that she needed additional physics and chemistry courses to be admitted to Hopkins. Reed chose to take them at MIT, which had started admitting women two decades earlier. The academics proved manageable, but the climate at a traditionally male school was an eye-opener. Reed learned to get to class early. Otherwise, as soon as she arrived, all the men stood up out of respect for her delicate nature. Yet those same male classmates had no qualms about stealing her laboratory equipment and sabotaging her experiments.[23] The thinly veiled harassment notwithstanding, Reed passed all of her courses with the highest marks available and promptly wrote to Welch. He replied, telling her she could enter Johns Hopkins Medical School the following October "without examination and without condition."[24] Reed's mother and other relatives were appalled with her choice of a career, presumably because medicine at the time was a low-status profession that often involved working with society's poorest members. Her elderly aunts refused to acknowledge that Reed was attending medical school, instead insisting when acquaintances asked about her that she had gone "south for the winter."[25]

Medical School at Hopkins: Confronting Sexism and Bias

Reed entered Hopkins as a member of the class of 1900 (figure 8.1). Her first day in Baltimore was one she would always remember. As Reed later recounted in her memoir, she had just stepped off the horse-drawn tram bound for Johns Hopkins Hospital when she noticed a stranger dressed in a morning coat and silk top hat who seemed to be following her as she walked briskly toward the hospital. The stranger suddenly inquired:

FIGURE 8.1. *Reed, circa 1895, the year before she entered medical school at Hopkins.*

"Are you entering the medical school?"

Startled, but determined to keep her composure, Reed later recalled she "managed to gasp out" that attending the medical school was indeed her intention.

"Don't. Go home," the mysterious man replied.[26]

The blunt, unwelcoming advice was uttered as the stranger strode past her.

Fortunately for us and for medicine, Reed was undeterred. The next day she presented her credentials to a panel of senior faculty. Dean Welch chaired the panel, and to her amazement, she recognized one of the members of the panel as the very same man who the day before had told her to go home. After leaving the interview room she asked a man waiting outside, "Who was the gentleman sitting on the left of Dr. Welch?" He answered, "Why, that is the great Dr. Osler."[27]

Reed's startling encounter with Osler in the street neatly encapsulates the daunting obstacles that Victorian women faced, even at a progressive institution like Hopkins, when they chose to pursue a career in medicine. (Reed would go on to develop an excellent relationship with Osler, who became one of her mentors and strongest supporters at Hopkins.) The episode also reveals the reservations that even a forward-thinking physician like Osler held about a woman's prospects in the late nineteenth-century medical profession. It also speaks volumes about Reed's grit, poise, and determination. In a medical career that spanned five decades, she often had to call on all these qualities to confront the challenges of an age that devalued women.

Reed rose to the challenge. She and Florence Sabin, whom we mentioned earlier, both graduated with honors. Along the way, together with their female classmates, they faced many academic and social barriers, as well as overt sexism. Under Hopkins's new curricular model, medical students spent their first two years immersing themselves in basic science: anatomy and physiology in year 1 and pathology, bacteriology, and pharmacology in year 2. Clinical work started in the third year. Reed complained that anatomy, where each student was given their own "box of bones," was "the greatest bore."[28] It was not until her second year that Reed's interest in medicine caught fire. In addition to using a microscope in her pathological histology course to understand the changes at the cellular level that underlie diseases, she attended Welch's weekly lectures. Reed gave him rave reviews, writing: "The clarity, simplicity, and remarkable synthesis of his material—made the most perplexing and difficult problem seem simple. Always pointing out, however, how much needed still to be cleared up and how incomplete was our knowledge of apparently simple phenomena."[29] Her first two years at Hopkins were demanding, but Reed excelled in her courses.

Living in a single room in Miss Conway's boarding house, Reed, who had been outgoing and popular at Smith, was lonely. The problem was alleviated in her second year when she moved into a rented apartment with three other women, one a close friend from Smith. As for male companions, soon after arriving in Baltimore, Reed discovered that Charles Mendenhall, a childhood friend from Columbus, was studying physics at Johns Hopkins. At some point after she moved in with her roommates, Reed began accompanying him on long Sunday afternoon walks. Though Charles later claimed he was instantly smitten with Reed, she apparently considered their relationship only an enjoyable friendship.

Harder to tolerate than her social isolation as one of only a small minority of women at Hopkins was the persistent sexism she encountered from her male classmates and even senior physicians (figure 8.2). In one notorious episode, which Reed could not forget or forgive, she and one of her roommates decided to attend a meeting of the Hospital Medical Society.[30] They were the only women present. The speaker, a clinical professor of laryngology, was preparing to deliver a paper titled "The Physiological and Pathological Relationship between the Nose and Sexual Apparatus of Man."[31] As Reed described it, "From the start he dragged in the dirtiest stories I have ever heard, read, or imagined, and when he couldn't say it in English he quoted Latin from sources usually not open to the public."[32] Reed's knowledge of Latin, unfortunately, made even the crude phrases in Latin understandable to her. Sitting with her roommate in the front, Reed knew she couldn't—or wouldn't—flee. Decades later she recalled, "We sat just opposite the speaker and the chairman, so that the flushed, bestial face of Dr. MacKenzie [the speaker], his sly pleasure in making his nasty points, and I imagine the added fillip of doing his dirt before two young women, was evident."[33] When the talk ended, Reed and her roommate made their exit, with Reed recalling the "sea of leering, reddened faces."[34] Reed considered giving up on Hopkins and medicine, but soon recognized that decision would not change the ugly, demeaning male attitudes she had experienced. She was determined to be "strong enough to rise above such defilement."[35]

Vulgarity and disrespecting women as peers was one thing, but denying them the professional opportunities they had earned through superior performance was quite another. Reed fought fiercely when Hopkins Hospital superintendent Henry Hurd tried to rob her and Florence Sabin of the coveted internships under Osler in the Department of Medicine they had won by their high class ranks at graduation.[36] As Reed wrote in her memoir, "There was apparently quite a lot of bad feeling brought about by my being given [an internship in] medicine."[37] But knowing that she had the support of Dr. Osler and Dr. Welch, she left Baltimore to spend the summer with her family "with high hopes."[38]

FIGURE 8.2. *Reed at Hopkins. Her strength is evident in her posture and expression.*

Reed returned in August to find Florence Sabin distressed. Sabin had been summoned to the superintendent's office, where Hurd had informed her that since two females had been named as interns, one of them would have to treat male patients in the ward for Black patients (designated the "Negro ward," or "colored ward" at that time). Hurd told Sabin in no uncertain terms that "it was unheard of for a woman to be in charge of the Negro wards. It would end in disaster, it couldn't be done—he wouldn't stand for it."[39] He had apparently further suggested (and gotten the reticent Sabin's agreement) that she study anatomy for a year rather than take up the internship in medicine. Unwilling to sacrifice what both women had worked tirelessly to earn, Reed went to talk to the superintendent. In a shocking demonstration of how sexism and racism collided in that era, Hurd accused Reed of having "abnormal sex interests" and suggested "only my desire to satisfy sexual curiosity would allow me or any woman to take charge of a male ward."[40] As for the "negroes," Hurd went on, "The man interne was all that kept them and the colored orderlies from insulting the nurses and women students—or worse."[41] He pointedly asked her "about rectal and genital examinations, catheterization, and other unpleasant duties" she might have to perform.[42]

Finally, Reed's anger at Hurd's unvarnished sexism boiled over. After dismissing his insinuations, she replied, "Dr. Osler gave me this appointment and said that Dr. Sabin and I would share the white and colored wards. He evidently considered both of us capable of running these wards to his satisfaction. Until he returns, sir, I shall be the interne of the colored wards, and I shall do my best. If in October I find that I cannot successfully perform my duties, I shall tender Dr. Osler my resignation."[43] She rose and left the superintendent's office, but not before summing up her feelings, telling Hurd her decision to enter medicine was a difficult one and that she had encountered some of "the unpleasantnesses" she had been warned about, but that she had "waited four years to be treated unfairly, and the first insult I have received was from the Superintendent of the Johns Hopkins Hospital."[44]

Reed had been blunt with the superintendent about her ability to handle the duties of an intern on the ward for Black male patients. She knew she had better prove herself as good as her words (figure 8.3). Reed worked seven days a week, from 6:00 A.M. before the first shift of nurses arrived until after midnight, when the residents made their rounds.[45] "Something Dr. Hurd had said of a woman's being irresponsible, and not to be trusted to see things through, kept me at my post."[46] As the doctor in charge, she had more to worry about than just medical care. On one occasion, she used a pair of crutches to break up a bloody melee that broke out among patients, shouting, "Back to your own wards!" as she waded into the fray.[47] Patients of the day had a habit of sneaking their preferred weapon—usually a straight razor—into the hospital, and at least a half dozen lay on the floor bleeding.[48] Still wielding the crutches, Reed made her way to the phone and summoned help. When order was finally restored, she spent the rest of the day stitching up wounds. Most of the injured were too sick to discharge for rioting.

FIGURE 8.3. *Reed (far left) with nurses in what appears to be the Octagon Building.*

Tough and cool when she needed to be, Reed was still able to win the trust and affection of her Black patients. On one occasion, Reed was called to the operating room, where one of her former patients lay bleeding. He desperately needed surgery but would not consent unless the person he trusted most, the "little white lady doctor," agreed.[49] When Reed came to see him near the end, he pressed his wallet filled with bills on her, saying "he wanted her to have it."[50] Reed, of course, declined. After Reed had spent six months on the Black male ward, Florence Sabin, who initially had been cowed by Hurd into saying she wasn't able to handle those duties, asked to trade places. Reed went to oversee the white women's ward, but once again her example had blazed the trail for other women.

FIGURE 8.4. *The medical interns and residents (house staff) of the Johns Hopkins Hospital in 1901. Reed (seated near the center of the second row) is the only woman.*

Another episode on the Black wards sheds light on Reed's exacting standards and her relationship with chief of medicine William Osler. Despite his initial warning to Reed to "go home," Osler was impressed with Reed's ability and became a strong ally. He had been away from Hopkins when Reed's angry exchange with Hurd had occurred, but by the time he returned, Reed had the Black male wards well in hand. Osler complimented her on the work, and she took pride in his approval. As his intern, Reed would accompany him on rounds and even come in early to update the notes Osler would consult at the bedside. On one memorable Saturday, Osler asked Reed to report on her work in "the Negro wards." Reed replied that there had been six deaths due to pneumonia, prompting derisive laughter from the audience of fellow trainees gathered around Osler. Reed, who held herself and everyone else to a high standard, was mortified. Osler quickly came to her rescue, noting that many of Reed's Black patients (albeit in his racially biased view) typically had other severe underlying illnesses and that under the circumstances she had done exceptional work (figure 8.4).[51]

A Love Affair Never Forgotten

On New Year's Day 1901, Reed met a handsome young physician, recently returned from the customary year or two spent studying in Germany, at a reception for house staff in the rotunda of the hospital.[52] They talked about Germany, which Reed knew well from her time spent there with Miss Gunning. When the party ended, they moved on to continue their conversation in the reading room. Reed never revealed the identity of the Hopkins physician, referring to him as "A. J.," apparently not his real initials. We do know from her letters that A. J. had done impressive research despite his young age. Reed's account of the early stages of their romance, however, revealed she had serious doubts about A. J. She felt he was egotistical, arrogant, and often self-absorbed.[53] These misgivings would ultimately influence Reed's decision not only about marriage but also about her long-term future at Hopkins.

Pioneering Work in Pathology

Earlier we described how William Welch dashed Reed's hopes for a faculty position after she completed her pioneering research on Hodgkin disease. We started with this episode because it crystallizes the barriers that Reed and other women faced in the medical profession. But the prejudice and discrimination that Reed faced only adds significance to her achievements. Revisiting her work in greater detail and in the context of medical science at the time helps us put its originality in a fuller perspective.

As Reed was completing her internship in the spring of 1901, Welch, likely impressed with her performance and recognizing her remarkable intelligence and work ethic, offered her a university fellowship in pathology. Welch's laboratory was the most advanced in the country. He and his colleagues employed techniques used extensively in Germany, including examining tissues collected in autopsies under the microscope and conducting experiments on animals. In addition, as noted in the chapter on Welch, both men and women were "made welcome" in his laboratory "to share in the most stimulating medical companionship."[54] Welch welcomed Reed and set her to work on a problem that had recently sparked his interest. Her meticulously researched solution would cement her claim to a place in the history of medicine.

Hodgkin disease, characterized by enlargement of the lymph nodes and the spleen, had first been described in 1832 by Thomas Hodgkin, a lecturer in morbid anatomy at Guy's Hospital in London.[55] Autopsies were the main tool for investigating diseases at the time, and Hodgkin based his findings purely on "gross" observations with his naked eye. Few advances were made in understanding the condition or its causes until the Austrian pathologist Carl Sternberg published a paper on the etiology of the disease in 1898.[56] Based on microscopic examination of the tissues of patients with Hodgkin disease and the fact that a number of them exhibited signs of active tuberculosis, Sternberg, as we noted, had incorrectly concluded that Hodgkin disease was a particular variant of tuberculosis.

At twenty-eight and after years of medical training, Reed was delighted to finally have her own scientific problem to research. Reed was skeptical of Sternberg's hypothesis. She studied the disease from multiple angles. First, she examined Sternberg's hypothesis that tuberculosis caused Hodgkin disease. As we noted, none of the animals Reed injected with ground-up lymph nodes from patients with Hodgkin disease developed tuberculosis. Instead, Reed observed that patients with Hodgkin disease often did not react at all when inactivated tuberculosis ("tuberculin") was injected into their skin.[57] Since tuberculosis was so common at the time, some of the patients should have had a strong reaction. Reed's finding was the first to show the "immunologic incompetence" of patients with Hodgkin disease.

The use of microscopy in research was in its infancy in this era, but Welch had given Reed a microscope "fitted with the new compound lenses that corrected" for deficiencies in earlier lenses.[58] Reed embraced and quickly became an expert in this new technology. She carefully and methodically studied tissues from patients with Hodgkin disease using her new microscope. Her findings became the centerpiece for the conclusions she would draw about the fundamental nature of the disease.

Reed's article summarizing the results of her research, "On the Pathological Changes in Hodgkin's Disease, with Especial Reference to Its Relation to Tuberculosis," appeared in the 1902 edition of *Johns Hopkins Hospital Reports*.[59] She concluded in unambiguous terms that tuberculosis has no direct role in Hodgkin disease, and that, instead, patients with Hodgkin disease had a weakened immune system.[60]

Reed next turned to her microscopic findings and concluded that the growth has a diagnostic appearance at the cellular level. Her paper noted: "The large giant cells, which are the most striking feature of these specimens . . . vary from the size of two or three red blood corpuscles to cells twenty times this size . . . the nucleus is always large . . . it may be single or multiple. These giant cells . . . are peculiar to this growth, and are of great assistance in diagnosis"[61] (color plate 12 and figure 8.5). For more than a century, this telltale marker in Hodgkin disease patients has borne the name Reed-Sternberg cell. In a single year of research, Reed used a new technology to refute an incorrect hypothesis on the etiology of Hodgkin disease and succeeded in characterizing its hallmarks. Nonetheless, because she was a woman, the faculty appointment she believed she had earned eluded her.

There may well have been another reason why Welch denied Reed a faculty position. By this time, Welch was aware that Reed had become romantically involved with A. J., who was likely William G. MacCallum.[62] MacCallum was a rising star on the pathology faculty, and Welch may not have wanted to deal with the complications such an "office romance" might have held for his department.

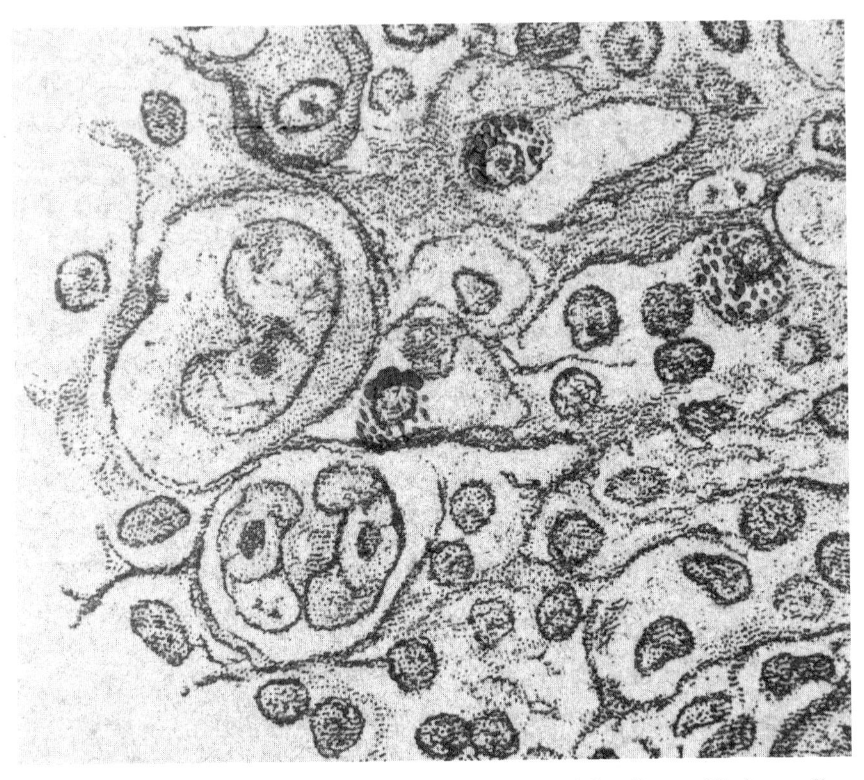

FIGURE 8.5. *Close-up of one of Reed's illustrations of Hodgkin disease. The large cell, now known as the Reed-Sternberg cell, that characterizes the disease is shown in the lower left. Note the two large, kidney bean–shaped nuclei that appear to face each other. (From D. Reed,* "On the Pathological Changes in Hodgkin's Disease, with Special Reference to Its Relation to Tuberculosis," *Johns Hopkins Hospital Reports.)*

Ironically, it has been suggested that Reed's love affair with A. J. ultimately contributed to the recognition of her pioneering research. In 1916, MacCallum wrote his authoritative *Textbook of Pathology*. In the section on Hodgkin disease, MacCallum repeatedly gives Reed credit for her discoveries, using phrases such as "Dr. Reed pointed out," and "Dr. Reed describes."[63] Reed's principal biographer writes, "She gave him her heart and in return he gave her international fame."[64] This interpretation should be taken with caution, as one could argue that it fits the all-too-common narrative that women succeed only because of their relationships with men. Certainly, Reed's sixty-four-page, single-author paper on Hodgkin disease stands on its own merits.

A New Beginning: Marriage and Pediatrics

With Welch's help, Reed secured a position as a staff physician at the new Babies Hospital in New York City. Her dreams of a career in pathology had been forever derailed. But with her characteristic mix of pragmatism and grit, Reed wrote, "May 30th, the very day I left Baltimore, was the lowest point of my life . . . On that day I turned my back on all I wanted most and started to make a new life for myself."[65] In part that pragmatism was driven by her continuing need to support her mother, whose financial situation was still precarious.

Because her appointment at Babies Hospital did not begin until January 1, 1903, Reed obtained an interim position at the New York Infirmary for Women and Children.[66] Reed was compassionate and caring but also demanding of herself and others. The experience of treating some of the city's poorest and sickest residents, as she had in Baltimore, left her torn between "sympathy with [the patients'] pitiful condition—deserved or not—and irritation with their feeble will or lack of mentality which got them into the situation and seemed bound to keep them there."[67] Established by Elizabeth Blackwell, the first woman to receive a medical degree in the United States, the New York Infirmary was staffed entirely by women. After six months at the New York Infirmary for Women and Children, Reed moved to Babies Hospital.

Reed's arrival at Babies Hospital on New Year's Day 1903 was a rude awakening. The physician in charge, Dr. L. Emmett Holt, at the time considered to be America's leading pediatrician, was vacationing in Florida when she arrived.[68] He had left her alone, with no instructions, in charge of all the babies in the hospital. Reed quickly discovered Holt did not keep medical charts on his patients, at best scribbling random notes on the back of their temperature charts. Reed methodically deduced their diagnoses by careful examination and by noting how the babies were grouped by ward.[69]

When Holt returned to the hospital, Reed encountered a short, dapper figure with piercing eyes and a brusque matter. In describing him she said, "he did not radiate kindliness."[70] On another occasion, Reed described Holt as "mercenary, a plagiarist [he apparently published the lecture notes of one of his nurses as his own work], and an unpleasant personality."[71] Both strong-willed people, Reed and Holt eventually developed a productive relationship, but it had none of the warm, collegial quality of her time with Osler. Reed did, however, learn much from Holt that would bolster her pediatric career. He was primarily interested in the problems of infectious disease and nutrition that were the mainstays of pediatric medicine in the early twentieth century. In 1894, Holt had authored *The Care and Feeding of Children: A Catechism for the Use of Mothers and Children's Nurses*, which became a bestseller.[72] The book went through seven editions and in 1946 was voted one of America's hundred most influential books. Holt covered the infant's first two years. The section on nutrition contained the following sound, if singsong, advice:

> What is the best infant's food?
> Mother's milk.
> What must every infant food contain?
> The same things which are in mother's milk.
> What are the symptoms which indicate that a child who is nursing is not nourished?
> It does not gain weight, cries frequently, sleeps irregularly . . .[73]

Reed would come to echo some of these ideas in her later writing on pediatric health. She spent nearly four years at Babies Hospital, a busy institution with nearly one thousand admissions a year. Holt depended on her for a wide range of clinical, surgical, and eventually pathological services.

Reed was at a crossroads during her years in New York—or more accurately a series of crossroads. She had chosen pediatrics more as a matter of necessity then choice. Her preference was research, notably research in pathology, where she remained current, but as a woman her path to holding an academic post was blocked. She also faced family and financial pressures.

The two converged when her sister Elizabeth died in 1903 of the chronic tuberculosis she and Reed had contracted as children. Only now did Reed learn that her sister's husband, whom Reed had always disliked, had squandered their money and ruined the business he had inherited.[74] The couple had three young children, whom Reed's spendthrift and frivolous mother, Grace, was incapable of raising. Despite being unmarried and early in her professional career, Reed felt compelled to assume responsibility for these youngsters. As Dawson notes in his biography, she would later even pay for her niece, also named Dorothy, to attend Smith College.[75]

Dorothy Reed married her longtime suitor Charles Elwood Mendenhall on Valentine's Day 1906, in Talcottville, New York, at the home where Reed's forbears had lived for generations. She chose the date because February 14 was her father's birthday. Mendenhall was by then a professor of physics at the University of Wisconsin. There is ample evidence that, despite Charles's "repressed, stifled nature," the Mendenhalls had a loving relationship and a stable family life, albeit one challenged by adversity and their very different ways of coping.[76] Both families disapproved of the match. Grace Reed, once thoroughly opposed to Reed's medical education, now lamented her daughter's decision to give up a promising career for a man with as little wealth, standing, and looks as Charles had. The Mendenhalls, for their part, ignored Reed's accomplishments. As she described it, "they never alluded to my being a physician or even introduced or spoke of my profession to me or anyone else."[77]

Early Years in Madison: Heartache and Hard-Won Lessons

After almost four years at Babies Hospital, Reed resigned to prepare for marriage and a new life with Charles. She became pregnant while the couple honeymooned in Italy. The coming months were consumed with setting up their new household in Madison and preparing for the arrival of their first child. Accustomed to the life of a professional woman, Reed found the transition to being a housewife challenging. But it is doubtful that she could have resumed working in those early years of marriage given the tragedies she soon faced.

When Reed went into labor for the first time in 1907, she was horrified by the care she received from the physician who was allegedly Madison's best obstetrician. The doctor hadn't properly washed his hands.[78] He tried to turn the baby in her womb (a maneuver known as a version) and ended up using forceps to deliver the infant in a breech position, inflicting a severe perineal tear on Reed in the process. The baby girl died of a cerebral hemorrhage later that day.[79] Reed had held beautiful, dark-haired Margaret on her in her lap for only moments but mourned her loss for years.

Reed became depressed after the baby's death and, in her own words, became "hard to live with," blaming herself for choosing the incompetent obstetrician.[80] Charles Mendenhall, though he had written Reed passionate love letters during their courtship, was self-contained and inwardly focused. He found it difficult to offer Reed the emotional support she needed. Nonetheless, in 1909, after a period of repair in their marriage, Reed gave birth to Richard, a beautiful, sweet-natured baby boy. But tragedy struck again. At the age of two, the couple's by then active, high-spirited toddler fell out of a window and died of a fractured skull. Dorothy and Charles had not wanted Richard to grow up as an only child. She had become pregnant again soon after delivering Richard. Thomas Mendenhall was born in 1910. He would go on to become the president of Smith College, where today Sabin-Reed Hall, a science building, is named in honor of Dorothy Reed and her fellow Smith graduate and Hopkins medical school classmate Florence Sabin.

Returning to Medicine with a Mission

The Mendenhalls's fourth child, John, was born in 1913. Reed had watched two of her babies die. She had endured other painful setbacks, including losing her mother, taking on added responsibility for her deceased sister's children, and being hospitalized after she was accidentally injected with carbolic acid by a careless nurse during a minor procedure to remove a mole from her face.[81] For weeks after the botched surgery, her "face was paralyzed, the skin sloughed, and saliva dripped out over her cheek."[82] Despite seemingly unending hardships, Reed gradually regained her balance and her desire to resume her career. Fighting hard against adversity was ingrained deeply in Reed's character, even when challenges temporarily overwhelmed her (figure 8.6).

FIGURE 8.6. *Charles and Dorothy Reed Mendenhall with their children, John (left) and Thomas (right), circa 1918.*

Her first opportunity came when Abby Marlatt, head of the Department of Home Economics at the University of Wisconsin, invited Reed to present a series of lectures on child health in the outlying counties surrounding Madison.[83] Her first lecture on nutrition was overly academic for her audience of farmwives. She quickly adjusted her style to take into account the information and advice most useful to rural women often dealing with poverty, hard physical labor, and sometimes abusive, liquor-fueled husbands. Most of these women breastfed their infants and had no one else to help them deliver their babies apart from poorly trained midwives. Reed empathized with them, but her real motivation in diving back into medicine was personal. "The tragic death of my first child, Margaret, from bad obstetrics in 1907, was the dominant factor in my interest in the chief function of women, the bearing and rearing of children. Most of my work came out of my agony and grief."[84] Reed went on to become the medical director of Wisconsin's first infant welfare clinic in 1915. A year later, she was appointed as a lecturer on child development in the Home Economics Department, though she never was invited to join the faculty of the medical school. The course she developed on sex hygiene for students was one of the first of its kind.

Charles Mendenhall was summoned to Washington, DC, to work for the army during World War I. Reed followed him, finding a position with the US Children's Bureau. Her work as a medical officer with the Children's Bureau continued even when she returned to Madison at the end of the war. Reed authored or coauthored several important pamphlets, including "Milk: The Indispensable Food for Children" and "What to Feed the Children."[85] Both were almost certainly influenced by her training more than a decade earlier with Holt at Babies Hospital.

Reed's best-known work for the Children's Bureau was "Midwifery in Denmark," first published in 1929. Here again, she was motivated by her tragic experience giving birth to her first baby. She compared infant and maternal mortality rates in Denmark to those in the United States and concluded that American mortality rates were higher because of unnecessary medical procedures. Reed advised that a well-trained midwife was preferable to a physician who could not get to the laboring mother in time, a situation that often prevailed in rural areas in this country. Reed was invited to testify about her research before a US congressional committee and became a strong advocate for natural childbirth and the Danish model.

Reed spread the gospel of safe and sound medical care for mothers and their babies at home in Wisconsin as well as at the national level. By 1934, reflecting in large measure Reed's influence, Madison had the lowest infant mortality rate in the United States among cities of comparable size.[86] Charles Mendenhall was stricken with pancreatic cancer that same year. At the age of sixty, Reed put aside her medical career to care for her husband, who died in 1935, a full three decades before Reed herself passed away in 1964.

A Life of Purpose and Intention

In chapter 1, we saw that Mary Elizabeth Garrett fought long and hard with President Gilman and the trustees at Hopkins to ensure that women would be admitted to the medical school on an equal footing with men. Less widely discussed is the eleventh article of her will.[87] It directed that much of her fortune would go to the School of Medicine *if* Hopkins provided equal opportunities for women faculty. Dorothy Reed Mendenhall was a prime example of the type of female physician-scientists Garrett was advocating for, but the provision in Garrett's will was never invoked. Garrett died before her close friend and sole heir M. Carey Thomas, leaving Garrett's estate to be disposed of as Thomas wished. Thomas lived lavishly, and Garrett's money was spent before it could be used as a strategic instrument to ensure equity for female faculty at Hopkins.[88]

When we consider Dorothy Reed Mendenhall's life, it is tempting to focus on the would-have-beens and could-have-beens because she encountered so much discrimination and adversity. While we need to acknowledge the indisputable reality of the obstacles she faced, it is even more important to highlight the extraordinary inner qualities that enabled her to accomplish so much—more, in fact, than most men who faced far more favorable odds.

Reed (figure 8.7) was highly intelligent, as her academic record at Hopkins and her groundbreaking research on Hodgkin disease amply demonstrated. She was brave, often in ways that only a person with that rare ability to see the possibilities beyond daunting obstacles could be brave. After enduring losing a newborn daughter and then a two-year-old son, Reed overcame overwhelming depression and resumed a career, bringing science to the medical care of mothers and children. And, of course, Reed was—and had to be—possessed of unusual strength of will. How else could she have resisted her family's relentless pressure to forgo a higher education and a career? And why else would she have taken on the burden of supporting her sister's children on top of her demanding responsibilities as a young physician? Hers was a life of purpose and intention from which we can all learn.

FIGURE 8.7. *Bust of Dorothy Reed by Joy Buba, 1966–1967. The bust is located in Sabin-Reed Hall, Smith College.*

9

HELEN TAUSSIG

The Founder of Pediatric Cardiology

Helen Taussig entered Johns Hopkins Medical School in 1924, twenty-eight years after Dorothy Reed Mendenhall enrolled. Yet Taussig faced the same barriers that Reed had encountered a quarter century earlier. Taussig was a Phi Beta Kappa graduate of the University of California at Berkeley and the daughter of a prominent Harvard economist. By the time she arrived at Hopkins, she had already completed original research on the physiology of heart muscle cell contractions. Despite these seeming advantages of education and accomplishment, Taussig confronted dyslexia as a child and deafness later in life and had to overcome obstacles erected by colleagues who harbored attitudes mired in sexism. A problem solver with high standards, she defeated these obstacles by never losing sight of her goals and by recognizing that achieving them meant outperforming her male colleagues. She did it as well by mustering a fierce determination in the face of challenges, a determination likely fostered in her difficult childhood that sometimes could be taken for steeliness.

Not surprisingly, opinions were divided about the woman widely recognized as the founder of pediatric cardiology. Virtually everyone who knew Taussig agreed about her exceptional devotion to her very young patients—mainly infants and toddlers—whose abnormal heartbeats she literally felt with her fingertips. Many who worked with Taussig lauded her warmth and generosity, especially to her students. Other coworkers found her aloof and cerebral. In our preface, we noted William Osler's caution about the perils of biography. We will likely never fully understand Taussig. Nonetheless, her own reflections are revealing. Toward the end of her career, Taussig was inducted into the National Academy of Sciences. Asked to reflect on the groundbreaking blue baby operation, which surgeon Alfred Blalock performed for the first time while Taussig stood beside the operating table and watched, she wrote simply, "The idea was mine. It grew out of careful observation of cyanotic infants."[1]

Her clear-eyed, matter-of-fact statement about this turning point in the treatment of babies with congenital heart disease is indisputably true and typical of her frankness. Our purpose in this chapter is to establish the pivotal achievements in Taussig's notable career and to shed light on a life story that equipped this remarkably forward-thinking physician with the inner resources to accomplish so much in a man's world.

Entering Uncharted Territory

By 1951, hundreds of children suffering from a congenital heart defect that left them with poorly oxygenated blood had undergone the Blalock-Taussig operation at Johns Hopkins. Despite the daunting challenges of performing delicate and extremely dangerous surgery on tiny and extremely ill patients, the mortality rate was only 10 percent. These children characteristically exhibited a bluish skin color (called cyanosis), clubbed fingers and toes, and breathlessness or fainting after even minor exertion—all the result of lack of oxygen in the blood, known as hypoxemia. Alfred Blalock, with guidance from his talented surgical technician Vivien Thomas, performed the first Blalock-Taussig operation in late 1944, and soon a steady stream of young patients was coming to Hopkins for the lifesaving surgery. The operation gave the vast majority of these hypoxemic children (so-called "blue babies") a renewed lease on life by establishing a new route for blood to travel from the heart to the lungs. Taussig's own path to conceiving the daring operation on tiny patients was long and arduous.

When she graduated from Hopkins medical school in 1927, Helen Taussig ran into the same roadblocks that had ultimately caused Dorothy Reed to leave Hopkins. Taussig had applied for the coveted medical internship—the position Reed had battled hospital superintendent Henry Hurd to keep. Taussig was denied the internship. Another female graduate with fractionally higher grades had already been accepted. The hospital would permit only one woman intern on the medicine service. Taussig instead accepted a research fellowship and spent the next year in Dr. E. P. Carter's adult cardiac clinic doing physiological research on the heart. In this short period, she published two original papers.

After her fellowship training, Taussig was invited by Edwards Park, the new chief of pediatrics at Hopkins, to become an intern in pediatrics at the Harriet Lane Home for chronically ill children, the first children's clinic in the United States affiliated with a medical school. Park, who had recently arrived at Hopkins from Yale, was a progressive clinical leader. He insisted on using X-ray and EKG technology at Harriet Lane and also purchased a fluoroscope, new technology that projected a rudimentary movie of a patient's beating heart on a fluorescent screen. Park preferred to appoint women to lead Harriet Lane's clinical activities, noting that men were prone to become impatient with pediatrics and abandon it for more high-profile services. By 1930, Park had named the exceptionally able and research-minded Taussig to head the Cardiac Clinic at Harriet Lane. Only thirty-two at the time, she was to hold the position for more than three decades until her retirement in 1963.

Reading Lips and "Listening" with Her Fingertips

Taussig soon found herself in charge of a bustling cardiac service, mostly filled with children afflicted with rheumatic fever. In those days before the widespread availability of antibiotics, streptococcal throat infections could quickly lead to heart damage, in many cases leaving patients with permanent heart muscle and heart valve scarring. Taussig instituted careful clinical examinations of her young patients, including taking blood pressures in both arms (since these could vary in cardiac cases), and employing X-rays, EKGs, and fluoroscopy. (The latter technique relied on patients swallowing a thick, unpleasant-tasting pudding with barium mixed in.) As Osler had done decades earlier, Taussig correlated her clinical observations with the patient's anatomy and pathology at autopsy.[2]

Park insisted that Taussig also learn about congenital heart malformations. Although she resisted at first, fearing that she would be trapped in a narrow, dead-end specialty, over time her interest grew. As she later wrote, "the luckiest thing in the world for me was that penicillin came along" to treat rheumatic heart disease, which previously had consumed so much of her time.[3] Hopeless as their condition seemed, Taussig often fed her tiny patients with congenital heart defects the bitter-flavored barium pudding and ran the fluoroscope herself. These fluoroscopic examinations enabled her to "diagnose the common cyanotic malformations by changes in the size and shape of the heart," and her increasing ability to establish these diagnoses reinforced her commitment to understanding congenital heart disease (figure 9.1).[4] She became fiercely devoted to her patients, referring to these children with complex heart defects as the "crossword puzzles of Harriet Lane."[5] Tragically, many of their puzzling abnormalities were only fully identified at autopsy. Taussig began her work in an era in when cardiology was barely recognized as a specialty and operating on the heart of an adult, much less a child, was considered far too risky. All she could do was to make a diagnosis and, all too often, watch a child die.

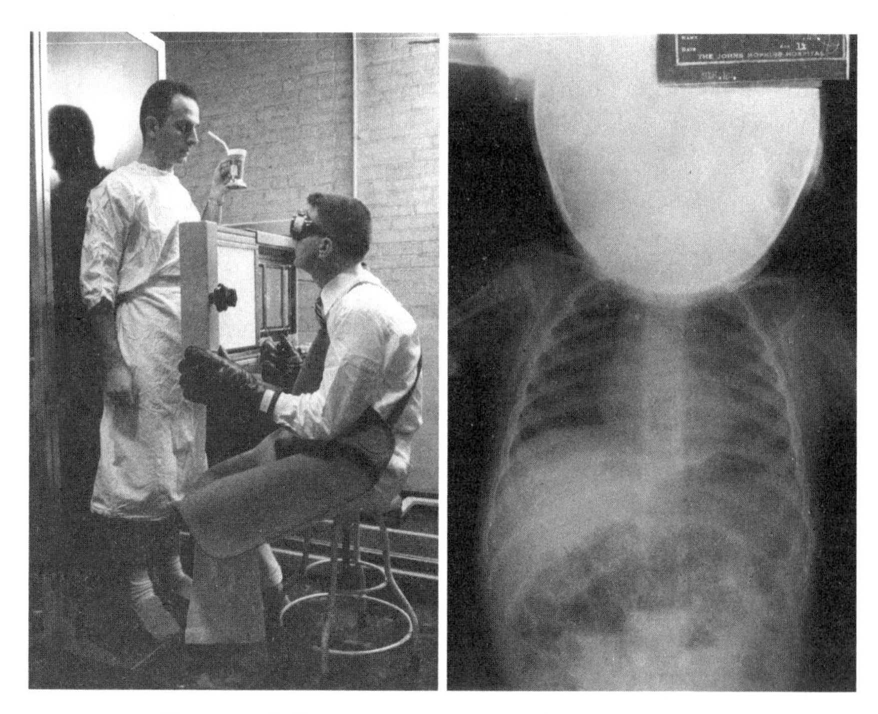

FIGURE 9.1. *Fluoroscopy (left) with man standing (1956). Chest X-ray (right) of an infant with the tetralogy of Fallot abnormality that produced the blue baby syndrome (1946). Taussig used both technologies to diagnose children with congenital heart disease.*

Taussig's challenges in running a busy cardiac service at Harriet Lane and learning to diagnose complex cardiac abnormalities were compounded by her own growing physical limitations. Her hearing had been impaired since childhood, the result of a case of whooping cough contracted when she was an infant. As Taussig neared the end of medical school, her hearing loss became severe. Voices, sounds around her, and, most critically for a cardiac specialist, the often-subtle murmurs of her patients' hearts grew faint. She started sitting in the front at conferences, learned to read lips, and used hearing aids (in those days, bulky vacuum-tube devices that she had to hang around her neck). For a time, she employed an electrically amplified stethoscope.[6] She gave up on the gadget due to poor sound quality, using it mostly when she wanted to share a patient's heart sounds with her students. Remarkably, Taussig taught herself to place her hands on a patient's chest and "listen" to heart sounds with her fingertips. She would later claim that her touch became a more sensitive diagnostic tool than any stethoscope (figure 9.2).

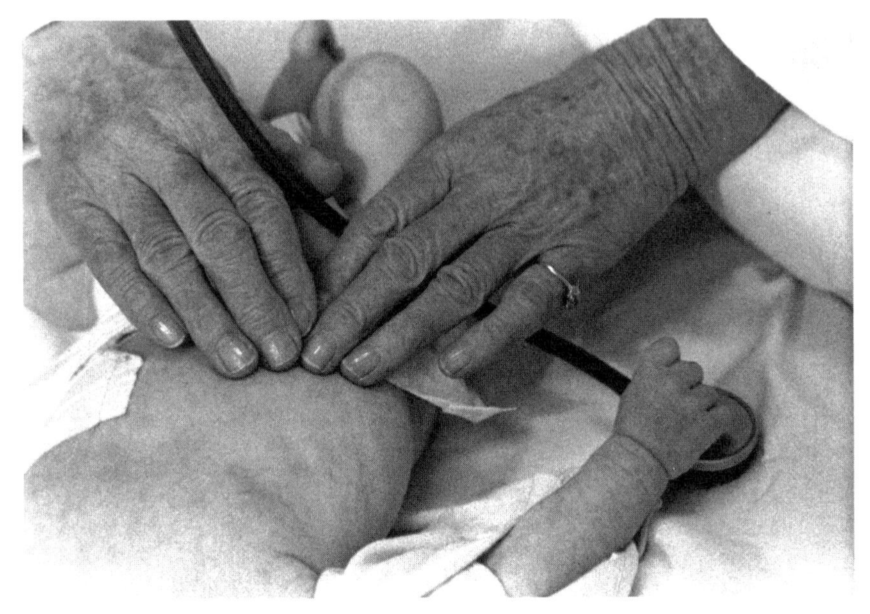

FIGURE 9.2. *Nearly deaf, Taussig would use her fingers rather than a stethoscope to "listen" to a baby's heartbeat. (Photo by Morton Todder.)*

Taussig's deafness undoubtedly affected her interactions with other physicians and perhaps contributed to her reputation for a certain aloofness. Colleagues who worked with her recalled that she had a tendency to either shout or whisper in conversation. Her deafness made it difficult for her to follow conversations in meetings. Numbers and symbols were particularly challenging, and she sometimes asked her coworkers to write them down. Whatever her challenges in communicating, Taussig remained unfazed, believing she had a job to do. As long as she was able to accurately sense her patients' heartbeats with her fingertips, her difficulty in making herself understood remained her colleagues' problem. Only when she was in her sixties did Taussig undergo the elective surgeries to improve her hearing that others had long urged on her.

From Blue to Pink: A Daring Leap of Imagination

Despite her initial resistance to studying congenital heart abnormalities, Taussig gradually became intrigued with the complicated problems they presented and began to discover patterns in the malformations. In particular, she came to recognize differences in the shape and size of the heart among patients who had, for example, transposition of the major arteries leading away from the heart and others who had an abnormality in the wall separating the two ventricle chambers. Then there were also children who had a condition called tetralogy of Fallot, a congenital abnormality characterized by four defects. Tetralogy, as it is known, includes a narrowing of the pulmonary artery (the artery leading from the heart to the lungs), a hole in the muscle between the left and right sides of the heart, an abnormal origin of the aorta (the main artery leading out of the heart to the entire body), and an enlarged right side of the heart. This combination of defects produces an insufficient flow of blood into the lungs. As a result, the babies are hypoxemic (in other words, their blood is not adequately oxygenated) and their skin is blue. These patients often died at a very young age.

Taussig came to understand that cyanotic infants died from a lack of blood flow to the lungs, not from heart failure. She later said, "As soon as I realized the cause of death I dreamed of being able to help these children."[7] She then made an astute, and deeply surprising, observation about cyanotic babies that would have far-reaching implications for her career and for the course of pediatric cardiology. Humans are born with a connection, a small blood vessel, between the aorta and pulmonary artery, the two major blood vessels leading out of the heart. This vessel is known as the ductus arteriosus. Its function in the womb is to reroute blood from the heart, bypassing the lungs, since a fetus receives its oxygen from the mother through the placenta. Normally, the ductus closes soon after birth and the infant's lungs take over the task of oxygenating the blood. Taussig noticed that some of her cyanotic patients, especially those with a tetralogy, died when the ductus closed. Paradoxically, those whose ductus did not close, a condition known as patent (or open) ductus, lived longer. After meticulously analyzing the data she had collected, Taussig concluded that in some of her patients a patent ductus arteriosus, or PDA, was allowing blood from the aorta to flow into the pulmonary artery, increasing the flow of blood to the lungs. What was in most circumstances a fatal combination of heart defects that deprived babies with tetralogy of life-sustaining oxygen was somehow compensated for by the PDA.

Taussig's logical, if entirely unprecedented, inference was that cyanotic babies with tetralogies might be saved if the beneficial oxygenating effects of a PDA could somehow be artificially replicated by surgically fashioning a vascular channel from the aorta to the pulmonary artery. No surgeon had ever attempted such a radical procedure, but the concept for the blue baby operation had been born. For the moment, all Taussig could do was to continue to document her observations about cyanotic children—and watch many of them die as their PDAs inexorably closed.

Things changed in 1939. Taussig read an article about Robert Gross, a surgeon at Boston Children's Hospital, who had operated to close a PDA in a child whose patent ductus was the only heart abnormality. She thought to herself, "What about turning that idea upside down—keeping the vessel open as a way to increase blood flow and help the cyanotic patient?"[8] Taussig traveled to Boston to meet with Gross. On the train there, she rehearsed all the reasons why opening a ductus would benefit tetralogy patients. Gross turned her down. As she described it, "It seemed pretty foolish to him to have me suggest he put a ductus in again . . . I think he thought it was one of the craziest things he'd heard in a long time."[9]

Refusing to Take No for an Answer

Faced with an eminent surgeon's flat-out refusal to consider an operation that would open a new path for blood to get to the lungs, many a physician would have called it quits. Certainly, there were compelling reasons not to push ahead. Surgery on the heart of a child, even surgery on the vessels close to the heart, was extremely risky. Some children with congenital heart defects survived well into adulthood despite impairments to their vigor and quality of life. Why roll the dice on an experimental procedure? Still, the grim spectacle of cyanotic babies who suffered oxygen deprivation that could leave them unconscious for a half hour or more motivated

Taussig to keep searching for a surgical solution to this heart abnormality (figure 9.3).

FIGURE 9.3. *Taussig in 1940. Her determination and grit are apparent.*

Alfred Blalock arrived from Vanderbilt as chief of surgery at Hopkins in 1941. A small, intense man who wore glasses and had swept-back hair, he was descended from southern aristocracy, tracing his roots on his mother's side to Jefferson Davis, president of the Confederacy. He seemed an unlikely surgical pioneer and an even more unlikely collaborator with Taussig, a reserved New Englander—unlikely, perhaps, except that Blalock, who had trained at Hopkins, had gone on to do remarkable work at Vanderbilt on the treatment of shock. That research would ultimately save thousands of wounded soldiers during World War II. Much of the research had been conducted using dogs as experimental subjects, in many cases tying off arteries leading from the heart to induce clinical shock.

Blalock brought with him from Vanderbilt Vivien Thomas, his talented longtime surgical technician. Thomas was tall, with a high forehead, a handsome face, and a habit of smoking a pipe. Thomas and Blalock were a study in contrasts. The white man with a deep drawl and the Black man born in a small riverbank town in Louisiana would occupy widely separated positions in the social and academic hierarchies at Hopkins. What they shared was a devotion to solving difficult surgical challenges.

As early as 1942, Blalock conducted the first-ever surgery at Hopkins to close a patent ductus. A veritable who's who of Hopkins physicians turned out to witness the cutting-edge surgery. Following the successful operation, Taussig took the opportunity to visit with Blalock. She told him:

> I stand in awe and admiration of your surgical skill but the truly great day will come when you will build a ductus for a cyanotic child who is dying from a lack of circulation to the lungs in contrast to tying off a ductus for a child who has a little too much blood going to his lungs.

To which Blalock replied:

> When that day comes this will seem like child's play.[10]

Perhaps having learned from her fruitless meeting with Gross several years earlier, Taussig was careful not to propose a specific surgical solution. When Blalock invited her to meet with him and Thomas in his laboratory,

Taussig merely referred to doing something for children with tetralogies, much like what happens when "a plumber changes [the] pipes around."[11]

Despite their very different temperaments, Taussig felt comfortable working with Blalock. She was ready to place the fate of her very ill babies in his hands. Still, before that could happen, countless hours of research in the dog laboratory, virtually all of it performed by Thomas, had to be conducted. Between 1942 and 1944, Thomas operated on hundreds of dogs. His task was to first create a blue baby-like condition in the animals, and then to try to correct the condition by anastomosing (or joining) the subclavian artery (the artery to the arm) to the pulmonary artery. In Thomas's words, even the initial task of simulating the conditions that existed in these congenitally defective hearts "would surely not be easy. Yet, to test any curative operation we first had to reproduce the condition in an experimental animal."[12]

Blalock's busy schedule as chief had allowed him to be directly involved in the laboratory experiments surgically joining the subclavian and pulmonary arteries just once—and then only as an observer. Nonetheless, by the fall of 1944, Taussig came to Blalock with an urgent case. Fifteen-month-old Eileen Saxon had no other chance for survival. She had weighed only nine pounds at eleven months and had spent most of the summer of 1944 in an oxygen tent at the Harriet Lane Home. Outside the tent, Eileen could barely drink without losing consciousness. Taussig had earlier diagnosed a tetralogy of Fallot that starved Eileen's fragile body of oxygen. Eileen was sent home from Harriet Lane in July when further treatment seemed futile. She was readmitted in October. By then it appeared she had only a short time to live. Years later, Eileen's mother described the agonizing choice she and her husband faced. Arriving at Hopkins one day in November, she recalled:

> Dr. Taussig said she wanted to introduce us to Dr. Blalock, who would perform the operation on Eileen if we consented. Dr. Blalock had soft but big hands, so I asked myself how he could operate on a little baby like Eileen with such big hands. But he was very patient and said it was a great risk and he didn't want to talk us into it. At the same time, he said it was Eileen's

> only chance. We thought about it and thought about it. But after all, we had no choice, so at last we said yes. They decided to operate on November 29 . . . Dr. Blalock was . . . trying something new, and Eileen was the first patient he had for it.[13]

Like Eileen's parents, Taussig and Blalock faced a stark choice. Taussig asked her surgical partner whether it was too risky to operate on a baby with such slim chances of survival. He replied, "You don't do a new operation on a good risk; you do a new operation on a patient who has no hope of survival without it."[14]

When she was wheeled into the operating room on November 29, 1944, Eileen Saxon's minute body was barely visible under the surgical drape. It foreshadowed the challenges Blalock would face rerouting the fragile blood vessels of such a tiny heart. Blalock insisted that Thomas stand by his side in the operating room, and during the surgery, he frequently asked Thomas for confirmation that his technique was correct. After an intense hour and a half of cutting and suturing, Blalock finally removed the last surgical clamp. As soon as he did, the anesthesiologist, Merel Harmel, called out, "The color is improving. Take a look."[15] The team then "leaned over . . . and looked at [her] face" and "saw the cherry-red color of her lips. It was astounding . . . really quite dramatic."[16]

A week later, Eileen stood up in her crib. At least for the moment, her life had been spared, and the pathway opened to thousands of other lifesaving operations on children with congenital heart defects. Taussig conceived the surgery, Thomas tested and perfected the procedure, and Blalock operated on Eileen. We will pick up the story of their intertwined roles in the blue baby operation in our next chapter.

In May 1945, after two more successful blue baby operations at Johns Hopkins, Taussig and Blalock published "The Surgical Treatment of Malformations of the Heart" in the *Journal of the American Medical Association.*[17] In a striking omission, Vivien Thomas was not included as a coauthor of the article or credited as one of the principal developers of the procedure that we would argue by all rights should be known as the Blalock-Thomas-Taussig operation.[18] Taussig won widespread recognition for her role in pioneering the blue baby operation. Her two-volume textbook, *Congenital*

Malformations of the Heart, published in 1947, immediately became the bible of pediatric cardiology.[19] She was appointed associate professor at Johns Hopkins School of Medicine in 1946 and belatedly named a full professor in 1959, only the second woman to hold the academic rank in the medical school's history. (The first was Dorothy Reed's classmate Florence Sabin, who, as we noted, faced her own prolonged struggle to achieve that rank.) More honors and accolades flowed Taussig's way, especially in the later stages of her career. In one of her proudest accomplishments, she taught and mentored an entire generation of pediatric cardiologists known as the "Knights of Taussig." She helped found the specialty as a distinct subdiscipline, and by the mid-1960s, the heads of pediatric cardiology at Columbia, Cornell, New York University, and Yale—all of them women—had trained under her.

The collaboration between Taussig and Blalock was immensely productive, saving the lives of hundreds, if not thousands, of children with congenital heart defects, but it was sometimes contentious, marked by the clash of two strong-willed personalities. For her part, Taussig was seldom reluctant to stand up to the famous surgeon. Of course, he was a man and she was a woman at an institution where men occupied all of the positions of authority. He was a professor and department chair; she held a lower rank. He was a surgeon and she a pediatrician at a time when surgeons held the power in academic medicine. Nonetheless, when it came to issues ranging from whose name was listed first on joint papers to running the clinical program, Taussig stood toe to toe with Blalock as an equal. We see this in a letter she wrote Blalock on October 12, 1951. Upset that one of Blalock's residents had taken over managing one of her patients and moved him to another ward, Taussig scolded the surgeon, writing, "Patients who are referred to me should be my . . . patients . . . I think this request is so right that I ought not to have to make it in writing to you."[20]

Privilege and Pain: Taussig's Early Life

Helen Brooke Taussig was born on May 24, 1898, in Cambridge, Massachusetts, the youngest of her family's four children. Taussig's father, Frank

William Taussig, was a brilliant Harvard economist who had advised President Woodrow Wilson on the economic provisions of the Treaty of Versailles that ended World War I. Her mother, Edith Thomas Guild, was one of the first graduates of Radcliffe College, the new women's division of Harvard. She held two degrees in botany and passed her love of nature on to her youngest child. Helen's father eventually became chair of Harvard's economics department and earned a very comfortable salary. In addition to owning a large house blocks from Harvard's campus, the family had a summer home on Cape Cod. Frank spent mornings there working on his academic writing and joined his family and guests for afternoons by the sea. With her busy clinical, research, and teaching duties at Harriet Lane, Taussig had a similarly disciplined work style. When visiting her summer home in Cape Cod, she invited guests to fix their own breakfasts and then joined them on the beach in the afternoon.

Despite the advantages conferred by her parents' first-rate educations and comfortable circumstances, Helen's early years weren't without their share of adversity. As a toddler, whooping cough left her with a permanent hearing deficit. Her mother died of tuberculosis when Helen was eleven.[21] Helen also struggled with severe dyslexia throughout her childhood. Her recollections of the shame and isolation her reading difficulties produced were vivid and seemed to have given Taussig a special empathy for children who had to cope with disabilities that set them apart. A poorly understood condition, readily mistaken for low intelligence or plain willfulness, dyslexia often exposed children to harsh treatment. Repeatedly forced to read aloud in class, Taussig was beset by stomach cramps and anxiety—sometimes to the point where she broke down sobbing and ran home. With her mother bedridden, it fell to Frank Taussig to patiently and laboriously help his daughter overcome her reading difficulties. She developed a lifelong bond with her father, describing him many years later in a private letter as someone who "had a very strong sense of what was morally right and he certainly did build it into us, and I myself am grateful to him for many things and he helped me in my work and in my writing."[22]

Taussig entered Radcliffe, her deceased mother's alma mater, in the fall of 1917. She was apparently disappointed with the all-women's college and, after a summer spent with her family in California, transferred to Berkeley, graduating with a BA degree in 1921. The idea of becoming a physician had "been 'mildly' appealing to her in high school and college," and Taussig next set her sights on a career in medicine.[23] As in the case of Dorothy Reed, had it not been for the courage and vision of Mary Elizabeth Garrett in insisting that Johns Hopkins School of Medicine accept women on an equal footing with men, Taussig probably would never have become a physician. Certainly, a male-dominated medical profession put obstacles in her path at every opportunity.

Harvard Medical School was out of the question. It would not admit women for another two decades. Helen's father suggested public health as a field with greater opportunity for women. When she met with Dr. Milton J. Rosenau, dean of Harvard's School for Public Health, he informed her that as a woman she would be permitted to attend classes but not earn a degree. Helen's forthright reply was, "Who [is] going to be such a fool as to spend two years studying medicine and two years more, and not get a degree?" Rosenau's answer was, "No one, I hope." She ended the interview, telling the dean, "I will not be the first to disappoint you."[24]

Despite Harvard's refusal to allow her to enroll, Taussig was permitted to study histology in the medical school. Even then, she had to sit in the back of the auditorium for lectures, and her microscope and slides were placed in a separate room so that she would not distract the men. Ironically, a professor advised her to cut her losses at Harvard and switch to Boston University's medical school, where she could at least complete a year of anatomy along with courses in physiology and pharmacology. It was there that Taussig, still barred from enrolling as a full-time student, completed her original research on heart muscle contraction (figure 9.4). Her anatomy professor, Dr. Alexander Begg, was dean of the medical school. Taking note of Taussig's talent and determination, he encouraged her to apply to Johns Hopkins, the only well-regarded medical school that admitted women on an equal basis. Taussig transferred to the Johns Hopkins School of Medicine in 1924, one of ten women in a cohort of seventy. She was finally on a path to a career in medicine, though her struggles with discrimination and prejudice were far from over.

FIGURE 9.4. *Taussig at the microscope, circa 1960. Although this photograph was taken later in her life, she used the microscope to make significant research advances as a student in Boston and during her fellowship at Hopkins.*

Later Career: Sounding the Alarm about Thalidomide

By the late 1940s, the Blalock-Taussig operation, or blue baby operation as it came to be known in the popular press, established the principle that surgery could be used to help children with congenital heart defects. More than one thousand of these operations were performed at Johns Hopkins in the late 1940s and early 1950s on children who were brought to Baltimore from all parts of the country and around the world by their hopeful but often desperate parents (figure 9.5). The procedure solidified Taussig's reputation as a pioneer in the emerging field of pediatric cardiology.

Yet, the breakthrough in treating young patients with heart abnormalities was bittersweet. Eileen Saxon, the first infant to undergo the surgery, who had shown such dramatic improvement in its immediate aftermath, died of her heart disease before she turned two. Later in her career, reflecting on Eileen's death and other painful losses she had experienced, Taussig wrote, "It's the clinical errors that keep you humble . . . You have your sadnesses as well as your successes. One reads all about the successful operation, but not about the unsuccessful ones, the sorrow and background of hard work."[25] If anything, the setbacks only heightened Taussig's determination to advance medical knowledge and help her patients.

Though she was known as a pediatric cardiologist, Taussig's concern for children and mothers took an unexpected turn in her later years at Hopkins. In 1961, when Taussig was sixty-three, one of her former fellows, Dr. Alois Beuren, visited her before returning home to his native Germany. He brought news of an alarming number of children recently born in that country with a rare but devastating birth defect known as phocomelia. The term was derived from the Greek words *phoke*, meaning seal, and *melos*, meaning limb. In infants born with this condition, the long bones in the arms or legs, and sometimes both, were missing or shortened. The most severely affected babies had rudimentary flipper-like fingers or hands that grew directly out of their shoulders. As Taussig later described it, "the infants' arms had almost completely failed to grow; their arms were so short that their hands extended almost directly from their shoulders."[26] Alarmed by Beuren's report and recognizing that West Germany's sharp rise in phocomelia cases had not yet been publicly disclosed in the United States, Taussig traveled to West Germany in February 1962 to investigate.

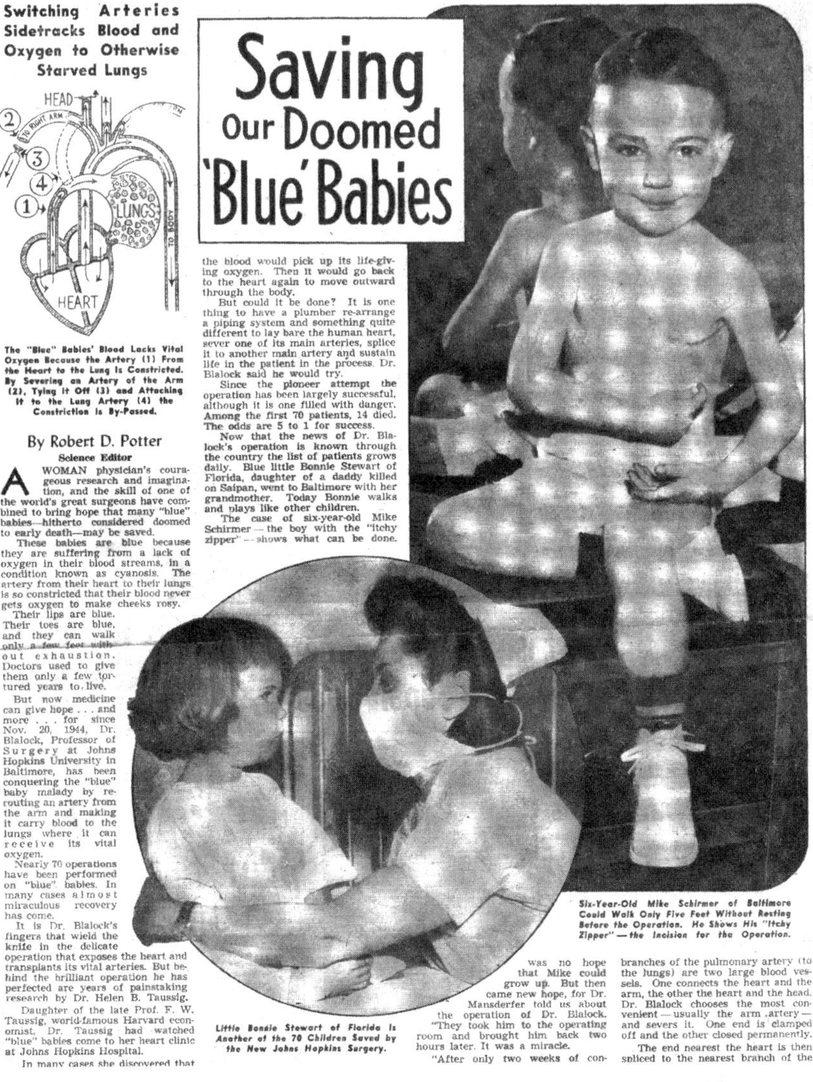

Switching Arteries Sidetracks Blood and Oxygen to Otherwise Starved Lungs

The "Blue" Babies' Blood Lacks Vital Oxygen Because the Artery (1) From the Heart to the Lung Is Constricted. By Severing an Artery of the Arm (2), Tying It Off (3) and Attaching It to the Lung Artery (4) the Constriction Is By-Passed.

Saving Our Doomed 'Blue' Babies

By Robert D. Potter
Science Editor

A WOMAN physician's courageous research and imagination, and the skill of one of the world's great surgeons have combined to bring hope that many "blue" babies—hitherto considered doomed to early death—may be saved.

These babies are blue because they are suffering from a lack of oxygen in their blood streams, in a condition known as cyanosis. The artery from their heart to their lungs is so constricted that their blood never gets oxygen to make cheeks rosy.

Their lips are blue. Their toes are blue, and they can walk only a few feet without exhaustion. Doctors used to give them only a few tortured years to live.

But now medicine can give hope . . . and more . . . for since Nov. 20, 1944, Dr. Blalock, Professor of Surgery at Johns Hopkins University in Baltimore, has been conquering the "blue" baby malady by rerouting an artery from the arm and making it carry blood to the lungs where it can receive its vital oxygen.

Nearly 70 operations have been performed on "blue" babies. In many cases almost miraculous recovery has come.

It is Dr. Blalock's fingers that wield the knife in the delicate operation that exposes the heart and transplants its vital arteries. But behind the brilliant operation he has perfected are years of painstaking research by Dr. Helen B. Taussig.

Daughter of the late Prof. F. W. Taussig, world-famous Harvard economist, Dr. Taussig had watched "blue" babies come to her heart clinic at Johns Hopkins Hospital.

In many cases she discovered that

the blood would pick up its life-giving oxygen. Then it would go back to the heart again to move outward through the body.

But could it be done? It is one thing to have a plumber re-arrange a piping system and something quite different to lay bare the human heart, sever one of its main arteries, splice it to another main artery and sustain life in the patient in the process. Dr. Blalock said he would try.

Since the pioneer attempt the operation has been largely successful, although it is one filled with danger. Among the first 70 patients, 14 died. The odds are 5 to 1 for success.

Now that the news of Dr. Blalock's operation is known through the country the list of patients grows daily. Blue little Bonnie Stewart of Florida, daughter of a daddy killed on Saipan, went to Baltimore with her grandmother. Today Bonnie walks and plays like other children.

The case of six-year-old Mike Schirmer—the boy with the "itchy zipper"—shows what can be done.

Six-Year-Old Mike Schirmer of Baltimore Could Walk Only Five Feet Without Resting Before the Operation. He Shows His "Itchy Zipper"—the Incision for the Operation.

Little Bonnie Stewart of Florida Is Another of the 70 Children Saved by the New Johns Hopkins Surgery.

was no hope that Mike could grow up. But then came new hope, for Dr. Mansderfer told us about the operation of Dr. Blalock.

"They took him to the operating room and brought him back two hours later. It was a miracle.

"After only two weeks of con-

branches of the pulmonary artery (to the lungs) are two large blood vessels. One connects the heart and the arm, the other the heart and the head. Dr. Blalock chooses the most convenient—usually the arm artery—and severs it. One end is clamped off and the other closed permanently.

The end nearest the heart is then spliced to the nearest branch of the

FIGURE 9.5. *A news story in the February 17, 1946, American Weekly about the blue baby operation. Stories such as this one spread the word about the surgery, and parents from around the world brought their children with congenital heart disease to Hopkins hoping for a cure.*

From the start, Taussig demonstrated a unique ability to voice her concerns as a pediatrician who cared for mothers and their children and to articulate these concerns in the larger context of the government's current drug approval standards. By the time she arrived in Europe, two different studies had identified thalidomide, marketed by the West German pharmaceutical company Grünenthal under the brand name Contergan, as the probable cause of the outbreak. Originally developed as an anticonvulsant, by the early 1960s, thalidomide was being widely used as a sleeping aid. It was distributed under various brand names not only in West Germany but also in several other countries. Thalidomide's apparent safety and especially the absence of fatal overdose potential made it popular in hospitals and psychiatric institutions as well as with consumers. Grünenthal frequently added thalidomide to aspirin and other drugs, promoting these over-the-counter remedies for conditions ranging from colds and coughs to headaches and asthma.

During her six weeks in Europe, Taussig visited the major clinics treating infants with phocomelia and interviewed officials from the major companies involved in the manufacturing and marketing of thalidomide, including executives from Grünenthal. They insisted that Contergan wasn't a causative agent in congenital limb defects, but Taussig came away with a strong suspicion about thalidomide. After visiting England, she was certain about its role in producing tragic birth defects. In the United States, the FDA had considered an application from Kansas City–based drug company Richardson-Merrell to market thalidomide under the brand name Kevadon, though approval had not been granted pending submission of further experimental data.

Speaking the Truth about the Drug Approval Process

Taussig returned to the United States in the spring of 1962. Her resolute actions to prevent the marketing and sale of thalidomide didn't reflect the mindset of a woman a year from retirement after a long and distinguished career. As a pediatrician, her deeply ingrained concern for the health and well-being of mothers and their babies made her quick to speak out against thalidomide and also to point to deeper problems with the approval of seemingly safe drugs. For Taussig, the thalidomide issue was about both the safety of a particular product and the *process* by which new drugs were approved and marketed. As she described it, "The necessity for this was demonstrated by the horrifying case of a British woman who had taken sleeping pills not knowing they were thalidomide. She gave birth to a deformed baby. She continued taking the pills, still not knowing what they were and gave birth to a second deformed child."[27]

Taussig sounded the alarm everywhere she could. In April 1962, she shared her experiences investigating thalidomide in a speech before the American College of Physicians. She also wrote letters about the hazards of inadequately tested drugs that were published in leading medical journals. Her message to her fellow doctors was simple: pay more attention to the unintended consequences of prescribing new drugs, and tighten the testing regimens that these pharmaceuticals must undergo.

In May 1962, Taussig was invited to testify before Senator Estes Kefauver's investigative Subcommittee on Antitrust and Monopoly when it held hearings about drug industry practices. Morris Fishbein, the influential editor of the *Journal of the American Medical Association*, was alerted to her upcoming appearance and publicly tried to undermine her testimony by questioning her qualifications to speak about drugs because she was trained as a physiologist and not a pharmacologist.[28] Taussig calmly offered the testimony she felt would sway a potentially skeptical audience. She recounted the tragic circumstances that had led to the widespread use of thalidomide abroad. She also made a point of showing pictures of the otherwise bright and engaging children who had been born with no arms or legs. In her concluding remarks, she took on Fishbein head-on, telling the committee: "Gentlemen, I am not a pharmacologist, and I may not even be a brilliant physiologist, but I think if any of you would have had a child or grandchild with this malformation, you'd do everything you could to prevent its happening again. That's why I'm here today."[29] Taussig's Senate appearance was persuasive. She recommended, and the committee, and eventually the full Congress, adopted a requirement that pharmaceutical packaging inserts describe potential adverse effects in the same type size as reported benefits.

Taussig's larger message was about the need to tighten regulatory oversight and to ensure that all new drugs were adequately tested for potential risks. It was only because of the quick and decisive response of an FDA medical officer, Frances Kelsey, that Richardson-Merrell withdrew their new drug application for thalidomide in March 1962. Taussig wrote:

> It happens that thalidomide-containing drugs did not reach the market in the US. This was because of a lucky combination of circumstances and the alertness of a staff physician at the Food and Drug Administration—not because of the existence of any legal requirements that the drugs might have failed to meet . . . I believe that it is essential to improve both the techniques for testing and the legal controls over the release of new drugs.[30]

Taussig also argued for stricter regulation of the process pharmaceutical companies used to place unapproved new drugs in the hands of doctors for the purposes of investigation. She recommended much tighter controls on drug company shipments and record-keeping and stringent standards to qualify physicians applying to investigate new drugs. In addition, she urged that "all the hazards of a given drug should be established before it is marketed."[31] In a sad footnote that underscored the urgency of Taussig's recommendations, Richardson-Merrell had distributed Kevadon, a drug combination containing thalidomide, to hundreds of US physicians for investigational use without adequate safety warnings. As a result, several "thalidomide babies," infants with missing or stunted limbs, were born in this country before the investigational use of the drug was stopped.[32]

The 1962 Kefauver-Harris Amendment to the Food, Drug, and Cosmetics Act gave the FDA authority to supervise the investigational use of new drugs. It also required positive consent from the FDA before a new drug could be sold to the public, and it empowered the agency to remove drugs from the market if new information revealed a public health danger. Taussig's testimony and efforts to spread the word about the thalidomide disaster and educate the public about the public health risks of existing drug approval laws were instrumental in passing this legislation.

Though hardly an ideologue, Taussig could be outspoken on a variety of issues, both medical and nonmedical. Her private papers in the Chesney Medical Archives at Hopkins even contain a letter she wrote on May 2, 1970, to President Richard Nixon, telling him, "I share with the young their concern over the extension of the war in Cambodia . . . it is forcing these young people to do what their conscience does not believe it is right to do . . . This, to me, is fundamentally wrong."[33]

A Busy Retirement and Belated Recognition

Taussig stepped down as physician-in-chief of the Cardiac Clinic at the Harriet Lane Home in 1963 after more than a third of a century as its head. She was sixty-five. Her retirement was more a refocusing, rather than a winding down, of her still considerable energies. She continued to be a full-time presence at the clinic. As professor emeritus, she kept seeing patients, making rounds, and attending conferences. Of the 129 scientific papers Taussig wrote during her career, forty-one were published after her retirement. "Now that I've retired," she quipped to one reporter, "I am much too busy to stop working."[34]

Being retired, at least officially, unleashed a flood of recognition and honors. In 1963, the March of Dimes Foundation established a fellowship for retired scientists carrying a $40,000 award. Taussig was named its first recipient and used part of the money to launch long-term follow-up studies of the children who had undergone the Blalock-Taussig operation between 1945 and 1951. The results were encouraging, both from a medical and a developmental perspective. Of the 432 patients operated on for tetralogy of Fallot who survived at least fifteen years after their surgery, 376 were still alive more than twenty years after their operation.[35] Even more heartening were Taussig's findings concerning the human outcomes of the blue baby operation. She found many high achievers—including doctors, lawyers, and other professionals—among patients who had the surgery as infants. The statistics were impressive: Of the 376 patients still living twenty years after surgery, 250 had married, 161 had children, 35 percent were college graduates, and nearly 70 percent were earning substantial sums of money.

Taussig took pains to emphasize that low arterial oxygenation in children born with congenital heart defects did not damage their mental capacities.

Taussig expanded her follow-up studies to include other congenital heart defects treated with surgical procedures and over the next decade published ten papers summarizing her long-term observations. Through it all, the flood of honors continued. In September 1964, she received the Presidential Medal of Freedom, the nation's top civilian award, from Lyndon Johnson. The following year, Taussig became the first female president of the American Heart Association. In 1965, apparently to her greater delight, she was the first woman and the first pediatrician elected president of the American Medical Association. The accolades Taussig received were not always readily accepted by her male colleagues. A Harvard surgeon's handwritten note in response to an article praising Taussig read, "Women are just physically incapable of enduring the long hours and hard work of my profession."[36]

In 1973, at the age of seventy-five, Taussig was inducted into the National Academy of Sciences, the country's most prestigious organization devoted to the advancement of scientific research. The recognition came belatedly. Alfred Blalock had been inducted twenty-seven years earlier for their work together on the blue baby operation. She had won the Lasker Award, widely regarded as a stepping-stone to the Nobel Prize in Medicine, two decades earlier. Closer to home, Hopkins has named one of the four colleges in the School of Medicine in her honor as well as the Helen B. Taussig Congenital Heart Disease Center.

A Life Balanced between Tenderness and Toughness

Several contrasting narratives emerge as we reflect on Taussig's life. One narrative, perhaps born of traditional gender expectations, portrays a female pediatrician who was deeply empathetic and loved the children she treated. Many of Taussig's peers "knew her as a kind woman, someone who cared deeply about them and the children to whom she ministered."[37] The way Taussig described the motivation for her work supports this narrative. In a speech she gave in 1953 after accepting an honorary degree from Western College for Women in Oxford, Ohio, she first reviewed the restorative effects of the blue baby operation on the children who underwent the life-saving surgery, their parents, and their communities. Taussig then turned her attention to the larger social implications of this medical advance. She asked rhetorically how progress is measured. Her answer was simple: "In medicine, progress is in alleviation and prevention of human suffering, the improvement of health and strength . . . Real progress implies some advancement towards the betterment of mankind."[38]

Another narrative sees Taussig as a pragmatist. There can be no doubt that Taussig was a role model for women's participation as full equals to men in the realms of work, society, and family life. She could be a forceful, even vocal, advocate for women when she needed to be. But Taussig was also a realist who took the world as she found it and who came of age long before the rise of the feminist movement. She had well-honed instincts for operating in a male-dominated world and good common sense about the best way to erase barriers that held women back.

In her handwritten notes for a lecture she delivered in the early 1970s to Phi Delta Gamma, a national fraternity, Taussig scribbled, "All prejudices breakdown if you can produce the goods"[39] (figure 9.6). Producing the goods and outperforming the men she worked with was how Helen Taussig arrived at the top of pediatric cardiology, a branch of medicine she virtually invented in the face of mostly skeptical male colleagues. Despite the difficulty of the medical challenges she faced with her very young and very fragile patients and in the face of discrimination from the men she worked with, Taussig never wavered in her determination to achieve the things she set out to do.

From a slightly different perspective, we might see Taussig as a vigorous, even combative, force of nature. When pressed and especially when it came to the medical care of her young patients, there was a steely intensity about Taussig that was unmistakable. When the time and the circumstances were right, she fought hard for herself and for others. She flatly refused Harvard's offer to study there with no hope of earning a degree. She had ignored Robert Gross's assertion that creating a new arterial path in patients with congenital heart defects was pointless. She battled Blalock over authorship priority on papers and control over the care of her patients. And, in the midst of the thalidomide controversy, she had bluntly and publicly challenged powerful pharmaceutical companies and argued forcefully for their regulation. Taussig seldom compromised where her principles or her priorities were concerned.

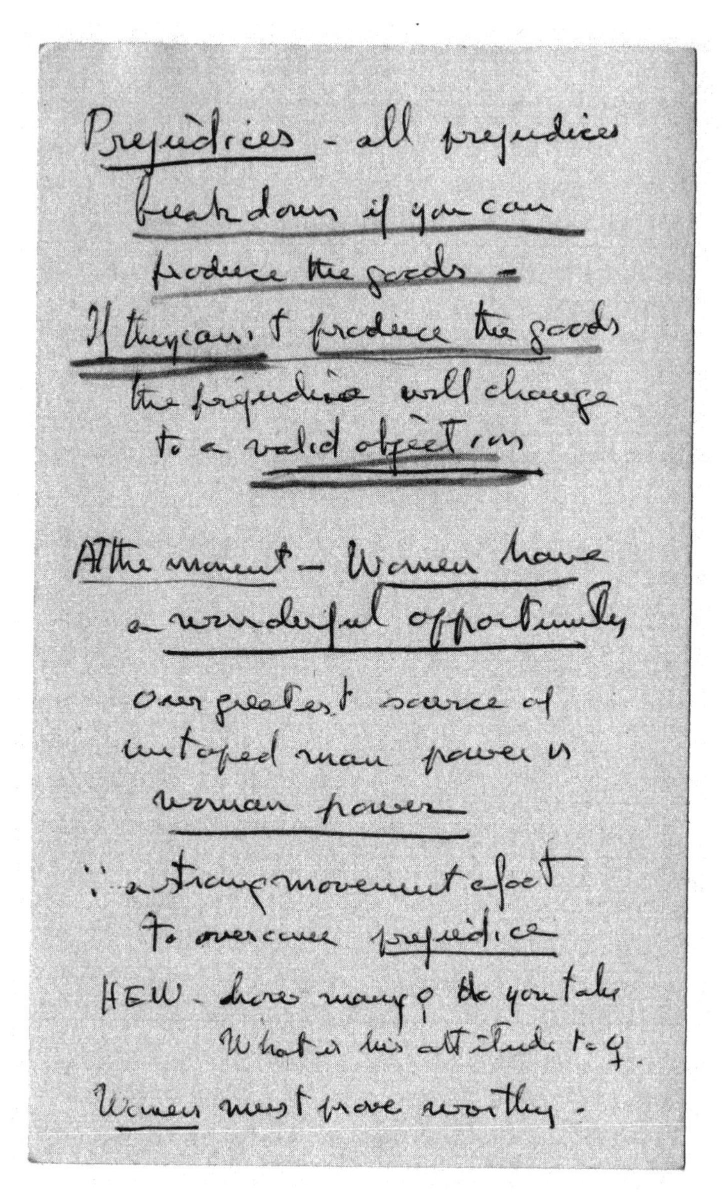

Prejudices - all prejudices
break down if you can
produce the goods -
If they can't produce the goods
the prejudice will change
to a valid objection

At the moment - Women have
a wonderful opportunity

Our greatest source of
untaped man power is
woman power

∴ a strong movement afoot
to overcome prejudice

HEW - how many ♀ do you take
What is his attitude to ♀.

Women must prove worthy.

FIGURE 9.6. *Taussig could be a pragmatist. She was accustomed to having to outperform men to get her fair share of credit. In notes for a talk to a group of students in the early 1970s, Taussig scribbled, "Prejudices—all prejudices break down if you can produce the goods."*

These competing narratives are reflected in the vastly different reactions to Jamie Wyeth's portrait of Taussig. In 1963, Taussig's former trainees commissioned the then seventeen-year-old Jamie, Andrew Wyeth's son, to paint her portrait. In preparation, Taussig's trainees sent the young Wyeth a photograph of Taussig. He couldn't stand the image, remarking, "She looked like a nice, middle-aged housewife with pearls."[40] Once they met, Wyeth came to know Taussig through long walks on the beach at her Cape Cod summer retreat and leisurely discussions over dinner.[41] The final portrait he put on canvas shows an older, intense Taussig with piercing eyes that seemed to take the measure of anyone who looked at her (color plate 13).

When the portrait was unveiled at a ceremony honoring Taussig on her retirement, the reaction was stunned but not silent. Many in the audience "began to cry . . . [and] were so startled by the portrait that the celebration ended abruptly."[42] They likely were shocked to see the woman they knew as having such "warmth and caring" painted in a dramatically different light.[43] A *New York Times* reporter writing about Wyeth's portrait years later called it "witchy" and said Taussig's friends described it as "evil."[44]

Not everyone disliked Wyeth's vision of Taussig. Langford Kidd, the director of pediatric cardiology at Hopkins in the 1970s and 1980s, said the painting "makes her look like a strong lady . . . Jamie Wyeth captured her essence . . . She wasn't a pushover. She was a very determined lady."[45] Another physician who knew Taussig well remarked, "She demanded excellence. If you didn't put out excellence, she let you know her displeasure."[46] A steely intensity, one that insisted on the best work from herself and from others, was central to Taussig's nature. That is not to say that the image we see in figure 9.7 of her tenderly examining a patient is somehow false. Caring and compassion, especially for children, were also deeply ingrained in Taussig's being. What Wyatt achieved in his unsettling painting was to give us a more complete, and in some ways more compelling, image of Taussig's character.

Wyeth himself described how Taussig posed for the portrait: "She would just stare right at me intently with those blue eyes. It was so amazing. I don't think she had a care in the least how she looked . . . I love the idea that everything was secondary to her work. I was trying to show Dr. Taussig's steely mind and questioning attitude. Not a Pollyanna woman but a very intense woman."[47] As her biographer Joyce Baldwin described it, the portrait "startles viewers with its strength."[48]

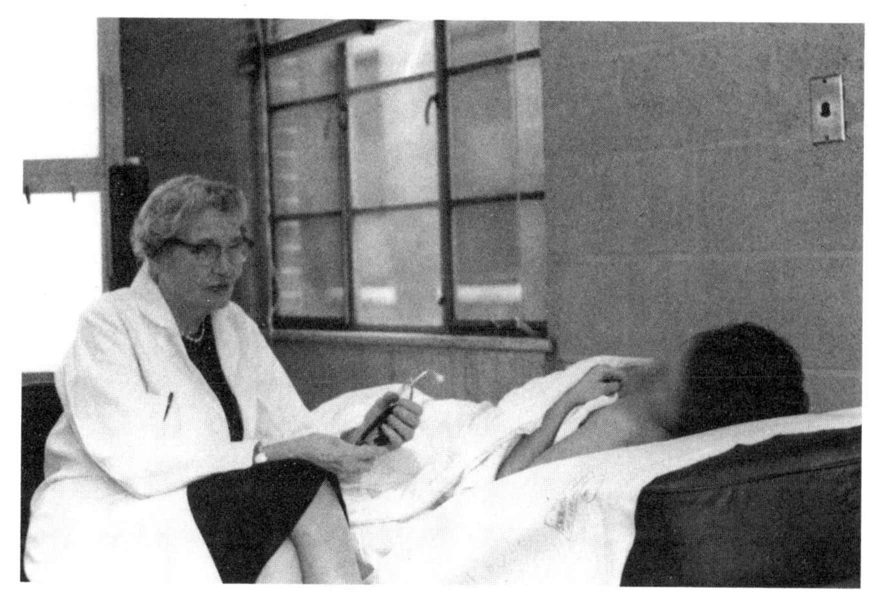

FIGURE 9.7. *Taussig was a caring pediatrician.*

Which of the competing narratives of Taussig is correct? Part of the charm of biography is that they can all be. Still, it is tempting to ask how Taussig saw Wyeth's portrait. Did it capture her true self? Unfortunately, we find little help answering this question. In her characteristic plainspoken manner, Taussig simply said, "I don't like it. I don't think many people do like their own portrait."[49] Whatever her private thoughts, for many of us, Wyeth's portrait allows us to see Taussig in a new and striking light, with all the intense drive and probing intellectual power she brought to her life.

Helen Taussig died four days before her eighty-eighth birthday, on May 20, 1986. She was driving a group of friends to vote in an election when her car struck another. She was killed instantly. The last article Taussig wrote, "World Survey of the Common Cardiac Malformations: Developmental Error or Genetic Variant," appeared in the *American Journal of Cardiology* when she was eighty-four.

It is impossible to think about Taussig's distinguished career without recalling her brilliant and imaginative medical mind, her tireless pursuit of a better way of treating malformations of the heart—often in the face of daunting personal and professional challenges—and her fierce devotion to her young patients. She was a problem solver with progressive views on the role of women in society and someone who held herself and others to the highest standards. Taussig will always be remembered for the daring conceptual leap that turned blue babies pink and opened the path to lifesaving surgery on congenital heart defects.

10

VIVIEN THOMAS

Something the Lord Made

Experience teaches us that truth can be stranger than fiction—and sometimes more powerful. Of our ten men and women who helped reshape American medicine, only one, Vivien Thomas, has been the subject of a Hollywood movie. Starring the rapper Mos Def as Thomas and the late character actor Alan Rickman as Alfred Blalock, HBO's *Something the Lord Made* (2004) won a Primetime Emmy for Outstanding Made for Television Movie and the American Film Institute's TV Program of the

Year award. The film, which was also nominated for a Golden Globe, is set in Nashville and Baltimore and features an intriguing performance by Mary Stuart Masterson as a nearly deaf but fiercely determined Helen Taussig.

Something the Lord Made tells the compelling story of the often-contentious collaboration between a driven, occasionally profane white surgeon and a high school–educated Black laboratory technician, which revolutionized cardiac surgery and saved the lives of thousands of children born with congenital heart defects. But for all its diligent efforts to re-create scenes and settings, this acclaimed, eminently watchable movie can never convey the complicated, sometimes fraught relationship between Blalock and Thomas or the blatant racism that Thomas confronted while leaving an indelible mark on heart surgery. Here, we trace as faithfully as we can the pivotal events in their relationship as well as the social milieu of Jim Crow in which both men operated.

Growing Up in Nashville

Vivien Theodore Thomas was born on August 29, 1910, in Lake Providence, Louisiana, a small town in the northeastern corner of the state, not far from the banks of the Mississippi River. Tired of the repeated floods that made earning a living and raising children an uncertain proposition, his parents, William Maceo Thomas and Mary Eaton Thomas, moved their growing family to Nashville in 1912. The city's burgeoning economy and position at a crossroads of commerce in the South made it possible for Vivien's father, a master carpenter, to earn a good living and build a substantial house on an acre of land in one of the city's Black neighborhoods.

Though indisputably part of the segregated South, Nashville was unusual in several ways. Most of its Black residents were poor, but the city also boasted a thriving enclave of Black doctors, lawyers, and businessmen. Two of Nashville's banks were owned by African Americans. Nashville was also the home of Meharry Medical College, one of only a handful of medical schools in the United States (another was Howard University in Washington, DC) that routinely admitted Black students. Three major Protestant denominations, including the African American National Baptist Church, conducted their publishing operations in Nashville.

After attending nursery school at Fisk University, a historically Black institution located a short distance from his home, Thomas was enrolled in Nashville's segregated public school system. By his own account, he received a good education. According to his autobiography, "The teachers in the public schools were concerned and dedicated . . . By today's standards, there was an almost unbelievable cooperation between parents and teachers . . . Discipline was seldom a problem; it was taken care of at home."[1] (We are fortunate in being able to rely on Thomas's autobiography, written when he was in his seventies, for firsthand accounts of his life and career.) Thomas attended Pearl High School, graduating in 1929. He described the teaching at this all-Black secondary school as excellent, noting that his teachers "knew their subjects, and knew their students and knew how to teach."[2] By the time he graduated, Thomas had completed courses in biology, chemistry, and physics as well as general subjects.

Besides school, there was plenty of work to be done in his father's carpentry business. Though never interrupting Thomas's classes, from the time Vivien was thirteen, William Thomas required his son to show up on a job site when he had a project in progress. Gradually working his way up from simple hammering and sawing to putting in entire staircases and banisters, Vivien became an accomplished carpenter by the age of sixteen. He had no trouble finding employment at nearby Fisk after he graduated from high school when the university put out its usual call for extra workers to complete repairs to dormitories, classrooms, and other facilities while students were on summer recess.

About to turn nineteen and possessed of a solid education, excellent carpentry skills, and a strong work ethic, Thomas seemed well situated to pursue his ambition to attend Tennessee State Agricultural and Industrial College and then pursue a medical degree. His prospects only seemed to improve when his foreman at Fisk laid off the other carpenters on Thomas's seven-man summer crew but offered to keep Thomas, a good worker, on through Christmas. He had been saving most of his wages to pay for tuition, books, and living expenses, but wasn't certain he had enough money in the bank to pay for both semesters. Wanting to build up his college fund, Thomas accepted the offer to stay on, figuring he would register at Tennessee State at midyear.

The Depression changed everything. A plummeting stock market and the collapse of the construction business forced Fisk to terminate his job in October 1929. His savings were wiped out when his bank went under. It was well past the deadline for registering at Tennessee State, even if he wasn't broke. Thomas knew he had to find another line of work—and fast. He asked Charles Manlove, a childhood friend who worked in the Bacteriology Department at Vanderbilt University, whether he knew of any jobs there. Manlove told him there was a Dr. Blalock in the medical school who had an opening, but "the guy was 'hell' to get along with and [he] didn't think I'd be able to work with him."[3] The next morning, February 10, 1930, Thomas went to see to Blalock at the medical school.

The two men sat talking on stools in Blalock's experimental surgery laboratory with the quiet but intense surgeon smoking and sipping a Coca-Cola (though not offering one to Thomas). Despite his friend's warnings about Blalock, Thomas found him "cordial and polite,"[4] and his description of the experimental work sounded quite interesting. Blalock took pains to describe the type of employee he was looking for. In Thomas's words, Blalock told him:

> As time goes on, I'm getting more and more involved with patients and hospital duties. I want to carry on my research and laboratory work and I want someone in the laboratory whom I can teach to do anything I can do and maybe do things I can't do. There are a lot of things that haven't been done. I want someone who can get to the point that he can do things on his own even though I may not be around.[5]

Thomas was impressed with Blalock's patient and detailed explanations of his experiments and clearly sensed an opportunity to learn. His reservations were around the pay. The $12 a week Vanderbilt was offering was just over half the $20 he had been earning the previous summer as a carpenter. Nonetheless, with no other job prospects on the horizon and college now more an aspiration than a certainty, Thomas told Blalock he would take the job.

The Vanderbilt Years

On his first day with Blalock, Thomas assisted him as best he could in setting up one of his experiments on a dog, helping to administer the anesthesia intravenously and assembling the test apparatus. The main equipment consisted of a mercury-filled device for measuring blood pressure and a rotating drum for recording it. (In those days, there were no electronic devices for capturing experimental data.) Blalock, he recalled, "directed my every move."[6] As Blalock finished up the day's experiment, he said, "We would do another experiment the next day and that I could come in and put the animal to sleep and get it set up. I was speechless. During my employment interview, he had told me that he wanted someone he could teach to do things. Had I gotten my first lesson and was I now expected to perform?"[7] Indeed, he was. Over the next several months, Thomas became proficient in setting up Blalock's experiments and often conducting them on his own (figure 10.1). Thomas could never be certain at what point in a particular experiment Blalock would sweep in from his hospital duties. The only sure bet was that Blalock would be present when an animal was autopsied to determine the results of an experiment.

FIGURE 10.1. *Thomas in the research lab at Vanderbilt.*

Blalock's work in the Laboratory for Experimental Surgery at Vanderbilt in the early 1930s was focused on the problem of why the human body went into shock after a severe injury. In those days, more patients with severe trauma died from infection or shock than from any other causes. Blalock's hypothesis, proven in hundreds of experiments, many conducted by Thomas, was that shock was not produced by bodily toxins, as the prevailing view in the medical establishment held. Instead, Blalock argued, shock was the result of a loss of blood and other fluids that led to circulatory collapse. Blalock's research paved the way to saving thousands of wounded soldiers in World War II and cemented his reputation as a rising star in American surgery.

Blalock's collaboration with Thomas was becoming increasingly productive. In the early months and years of their work together, the talented surgeon discovered in Thomas someone with exceptional technical proficiency. Over time, Thomas's medical knowledge and skills also grew rapidly. By 1931, Blalock's work on shock was taking a new, more physiological direction, attempting to explain the chemical changes in the body that occurred when traumatic injuries produced fluid loss. Dr. Joseph Beard, a postdoctoral fellow who worked in Blalock's laboratory, took it upon himself to teach Thomas about chemical processes well beyond the basics Thomas had learned in high school. Beard loaned Thomas his chemistry and physiology textbooks, and they often discussed the underlying rationale for new experiments Thomas was being asked to perform.

Working largely on his own and on time snatched from his busy laboratory schedule (Thomas and Blalock often worked until nine or ten o'clock at night), Thomas taught himself anatomy. He did it by embalming experimental animals with chemicals from the autopsy room and then dissecting them. "As time permitted," he wrote, "I would dissect these embalmed specimens, determining exact anatomical position of organs, their blood supply, and their relative position to each other."[8] Thomas's surgical skills also advanced during his years at Vanderbilt. Since he couldn't predict when Blalock would show up to participate in an experiment, Thomas often ended up conducting the entire procedure himself. His skills in vascular surgery, which would later become critical in developing the techniques for the blue baby operation, took major strides in 1935 when Blalock began investigating the effects on blood pressure of constricting the renal artery. The experiment first required a complicated and delicate operation to transplant the kidney of the experimental animal to the neck area, where it would be more accessible. Sanford Levy, a new postdoctoral fellow, had by then joined Blalock's laboratory. Working together, the three men all honed their new vascular surgery skills, but, as Thomas reported decades later, "the brunt of the transplantation fell on Levy and me."[9]

Thomas was clearly on his way to fulfilling Blalock's fervent wish to have "someone in the laboratory whom I can teach to do anything I can do and maybe do things I can't do."[10] His relationship with Blalock was constructive, but as a Black man working for a white doctor in the South, hardly uncomplicated. Things came to a boil about two months after Thomas started working at Vanderbilt. In his autobiography, Thomas said he couldn't recall the exact cause of Blalock's displeasure, but that one day his new boss threw "a temper tantrum" and lit into him with profanity that "would have made the proverbial sailor proud of him."[11] Finished with his rant, Blalock went to the laboratory's walk-in refrigerator, grabbed a Coca-Cola, and left. Thomas had heard enough. He went to the locker room, changed into street clothes, and walked across the hall to the surgeon's office to voice his displeasure and to hand in his resignation. According to Thomas:

> [Blalock] actually acted surprised, as if nothing had happened. I told him that he could just pay me off, that I was trying but if it was going to be like this every time I made a mistake and I couldn't please him, my staying around would only cause trouble. I said that I had not been brought up to take or use the kind of language he had used across the hall. He apologized, saying that he had lost his temper, that he would watch his language, and asked me to go back to work.[12]

Thomas confronted Blalock about his behavior, and perhaps more subtly, communicated that he would not tolerate the dehumanizing remarks often directed at junior Black workers in the South by their white bosses. By doing so, Thomas laid the foundation for a new, more meaningful relationship with Blalock that, although often contentious, lasted until the surgeon's death. Thomas concluded the story, writing:

> Dr. Blalock kept his word for the next thirty-four years . . . We had occasional disagreements and sometimes almost heated discussions. But neither of us ever hesitated to let the other know, in a straightforward man-to-man manner, what he thought or how he felt, whether it concerned research or, in later years, the administration of the laboratory. In retrospect, I think this incident set the stage for what I consider our mutual respect throughout the years.[13]

Respect was crucially important to the intensely proud Thomas, but he couldn't pay rent or buy groceries with it. Salary was a constant source of friction between Blalock and Thomas. Thomas's job classification at Vanderbilt was "janitor," and he was paid roughly half of what he had been making as a carpenter. When Thomas reminded Blalock of his promise to raise his salary after several months, Blalock said he would look into the matter, but nothing happened. A frustrated Thomas finally talked to his former foreman at Fisk and agreed to return there as a carpenter at his previous wage of $20 a week. He told Blalock he was leaving. The surgeon, upset and desperate to keep Thomas, asked him what he would be making. Negotiations ensued, and Thomas agreed to $17.50 a week, about 50 percent more than his previous Vanderbilt salary. Pay remained a bone of contention between the two men, especially several years later when Thomas married and built a new home for Clara, his bride. The pattern persisted at Hopkins, where on several occasions Thomas threatened to leave. For a good part of his career at Hopkins, Thomas needed to moonlight as a bartender or waiter to supplement his technician's salary.

Despite their arguments over money and their different personal styles, by the late 1930s, Thomas's and Blalock's careers and professional futures had become increasingly intertwined. Blalock himself had matured at Vanderbilt. Known as a tennis player and "a Casanova" when he was a medical student at Hopkins, Blalock found a new seriousness of purpose after coming to Vanderbilt in 1925.[14] He had married the beautiful and socially prominent

Mary O'Bryan (adeptly played in *Something the Lord Made* by the talented Kyra Sedgwick) in 1930. The wedding was the high point of Nashville's society season that year. More importantly, Blalock's work on shock as head of Vanderbilt's experimental surgery lab had brought him widespread recognition in medical circles. He had recently been made a full professor at Vanderbilt Medical School. Blalock was clearly ready for bigger things.

In 1937, Blalock received an offer from Henry Ford Hospital in Detroit to become chief of surgery. After asking Thomas to consider moving with him to Detroit, Blalock subsequently turned down the job. Though the evidence is not clear-cut, it seems that Henry Ford Hospital, which Thomas's sister described as "lily white," refused to accept Blalock's assistant as part of a "package deal."[15] The long-awaited invitation from Blalock's alma mater to lead the surgical department at Johns Hopkins finally came in 1940. This time Blalock insisted that Thomas be included in the offer.

The Move to Hopkins

Departing Nashville for Baltimore posed a dilemma for Thomas and his young wife. On the one hand, it would mean leaving behind their newly built home and Thomas's close relations. Thomas's own family was growing. The couple now had two young daughters to raise and educate. On the other hand, Thomas enjoyed the medical research he was doing at Vanderbilt, and there was no prospect of similar employment there after Blalock left. In the end, as Thomas described it, "Believing that we didn't have much to lose, that we were young and hopefully had many years ahead, we made this decision to move to Baltimore and to Johns Hopkins."[16]

Thomas arrived in Baltimore in June 1941, a few days ahead of Blalock, who had purchased a home in Guilford, a wealthy enclave of gracious residences in the northern part of the city. Thomas's search for a rental apartment was difficult and quite different. Baltimore was in full wartime production mode, even though Pearl Harbor and the country's entry into World War II was still six months away. The influx of thousands of workers seeking employment in Baltimore's bustling port and booming defense industry had created a severe housing shortage. "Many of the apartments that bore 'for rent' signs," Thomas wrote, "could hardly be classified as fit for human habitation."[17]

After weeks of searching, Thomas found an apartment he considered "marginal living quarters" but decided to take it with the intention of finding something better for his family in the coming months. Accustomed to living in Nashville with its single-family homes, trees, and lawns, all of which were typical even in the city's segregated Black neighborhoods, Vivien found that "this type of residence was just about nonexistent, or rather unavailable to Negroes, in the Baltimore area."[18] If Thomas had expected the racial climate in Baltimore would be different from the South, he was sorely disappointed. Cleveland Bealmear, chair of the Baltimore Housing Authority at the time, said of the influx of African Americans into Baltimore during the war: "They are going to bring in here from North and South Carolina, Negro workers. They are going to send us scum."[19] Racist attitudes like these pervaded Baltimore's housing bureaucracy, and only compounded the struggles of Thomas and thousands of other newly arrived Blacks to find a decent place for their families to live.

Blalock's ambition when he arrived as chief of surgery at Hopkins was to restore the department's renown in surgical research. That activity was centered in the old Hunterian Building, about a block away from the hospital. When Blalock and Thomas made their first visit there, they were dismayed. The building was dark and dilapidated, and most of its equipment was antiquated. After accepting Blalock's invitation to repaint their own rooms since he would be spending so much time there, Thomas set about requisitioning the supplies and instruments they would be using. Thomas presented a list of what he and Blalock needed to the director of the laboratory. He soon sensed that there was something wrong—and not just with the order.

Days went by, the supplies were eventually delivered, but two crucial items were missing. Thomas let Blalock know about the shortage. A furious Blalock grabbed a sheet of paper from his desk, scrawled a note to the laboratory director, saying, "Get anything Vivien asks for for my work," and told Vivien to hand-deliver it to the director of the laboratory.[20] Though the director would pass Vivien frequently in the laboratory's corridors, the man never spoke to Thomas again.

The episode was Thomas's first, but far from his last, encounter with the often unspoken but pervasive racism that would greet him as a Black man at Hopkins. Orthopedic surgeon Koco Eaton, Thomas's nephew and a 1987 graduate of the School of Medicine, has pointed out that both the hospital and medical school at Johns Hopkins are steeped in tradition and organized in strict hierarchies. For decades, the dress code required male interns to wear white pants and a short white coat. Only when you achieved the rank of a senior resident did you graduate to wearing your own pants and a long white laboratory coat.

FIGURE 10.2. *Thomas in his white lab coat, circa 1970.*

As it turned out, when it came to a Black man, even hallowed traditions could be enlisted in the service of bigotry. When not scrubbed for surgery in the experimental laboratory, Thomas was accustomed to wearing his white coat, always bleached within an inch of its life and starched to show not the slightest wrinkle (figure 10.2). On one occasion shortly after

Thomas arrived at Hopkins, Blalock telephoned from his surgical office in the hospital and asked Thomas to walk over with his notes and provide an update on his progress. "I put everything in a manila folder, put on my coat, and headed for the hospital . . . The distance . . . to the Osler building is about a city block along the main corridor. As I passed people, some of them actually stopped in their tracks and stared at me . . . I had had a long white coat on. A Negro with a long white coat? Something unseen and unheard of at Hopkins!"[21] Thomas wrote that, on learning about the incident, Blalock "was amused and smiled."[22] Eaton suggests that his uncle "was a man with pride, but yet a lot of times that pride couldn't show. So I think it was very tough for him."[23]

Recreating the Blue Baby Condition in Dogs

In their initial meeting, Taussig had challenged Blalock and Thomas to change "the pipes around" and build a new arterial path from the heart to the lungs to save cyanotic babies who were dying from insufficiently oxygenated blood. The challenge was certainly a technical mountain for Blalock and Thomas to scale. Surgery on the heart had rarely been attempted. Yet, the problem also represented an answer to Blalock's search for a major new surgical research project after his pioneering work at Vanderbilt on shock.

Before Blalock and Thomas could even consider a surgical solution, they first had to create a blue baby–like condition in the dog laboratory. With Blalock preoccupied with his administrative and teaching duties, the work fell largely on Thomas. He wrote: "Dr. Taussig had accumulated a sizable collection of these congenitally defective hearts in the museum in the Pathology Building. I spent hours and days poring over, examining and studying these preserved specimens trying to figure out if and how it would ever be possible to create such a model to study and to attempt surgical treatment."[24]

As described in the previous chapter on Taussig, tetralogy of Fallot, the most common cause of cyanosis in babies, is a combination of four different congenital abnormalities. Thomas considered three of them too complex and deeply buried in the heart to simulate in dogs. Only stenosis, or constriction, of the pulmonary artery supplying blood from the heart to the lungs, offered any prospect of being replicated. Even there, Thomas struggled to recreate the condition. His repeated and laborious efforts to tie off the delicate pulmonary artery, first with surgical linen, then with umbilical tape used in the delivery room, and finally with rubber tubing, all failed to produce the desired reduction in blood flow to the lungs. Finally, in a meeting with Blalock to discuss their progress, the surgeon proposed removing selected lobes of the lung combined with an anastomosis, or joining, of a branch of the right pulmonary artery to the right pulmonary vein, thus creating a blind loop and bypassing the flow of blood to the lung on that side. When Thomas returned to the dog laboratory and attempted the new procedure, it resulted in "reducing the amount of effective lung tissue for oxygenation" and mimicked the main complication of tetralogy of Fallot.[25] The path had been cleared to devising a surgical solution to a major congenital heart defect.

Once a blue baby–like condition had been created in experimental animals, the next step was to, in Taussig's words, find the right way to change "the pipes around" to correct the problem. For this, Blalock and Thomas connected (or, again in surgical terms, anastomosed) the subclavian artery (the artery supplying oxygenated blood to the arm) to the pulmonary artery. Blalock had done precisely that at Vanderbilt in 1938 in an unsuccessful attempt to produce pulmonary hypertension (high blood pressure in the lungs) experimentally. Of course, before the solution could be tried in humans, it had to be tested and retested in the dog laboratory. Once again, most of the work landed on Thomas.

Anna: The "Patient" Who Made It All Possible

As you walk through the corridors of the Johns Hopkins School of Medicine, you are greeted by dignified portraits of notable men and women in the institution's history. The distinguished figure gazing down on you on the sixth floor of the Physiology Building is equally poised, but you'll notice she has furry ears and a noticeably wet, black nose (figure 10.3). In 1943, Vivien Thomas and the surgical lab team performed the first successful blue baby operation on Anna, a full-grown dog whose exact age and pedigree were not known.[26] She was the first dog to survive the operation (or the Blalock-Taussig shunt, as it came to be known) and is rightly celebrated for her contribution to medical history.

FIGURE 10.3. *The Blalock-Taussig operation was performed on a dog, Anna, before it was tried on humans. (*Anna the Dog *by DeNyse W. Turner Pinkerton, 1951, oil on canvas.)*

Anna not only made it through the surgery, but she thrived, becoming something of a celebrity at Hopkins and in the wider Baltimore community. Postoperatively, Anna lived for another fourteen years in the Hunterian Laboratory, where she was a favorite with staff and visitors. When young patients who had undergone the blue baby operation returned for follow-up visits with Blalock and Taussig, they routinely visited Anna. Photographs of the children with the celebrated dog appeared in newspapers and magazines. Anna broke into the movies in 1950, when her story was told in a film made for school audiences and other animal-loving groups. In 1951, Anna's portrait was painted by DeNyse W. Turner Pinkerton, a locally renowned artist, and presented to Johns Hopkins Hospital. Alex Haller, one of Blalock's surgical residents who went on to become head of pediatric surgery at Johns Hopkins, has called Anna "a symbol of humane treatment of animals who were used in research."[27]

Putting Blalock and Thomas's Research to the Ultimate Test

By late November 1944, the infant Eileen Saxon's condition had deteriorated to the point where Taussig decided that surgery could not be delayed. She went to Blalock and insisted that he try. The surgery was scheduled for November 29, and Blalock and Thomas accelerated their preparations. The surgeon told Thomas that he wanted to assist him in performing a surgery to connect the subclavian artery to the pulmonary artery in a dog "and then do one or two with my [Thomas's] assistance."[28] Blalock appeared in the dog laboratory at the appointed time and assisted Thomas. For whatever reason, perhaps due to his heavy clinical and administrative schedule as chief of surgery, Blalock never returned to the dog laboratory to practice the surgery as the lead surgeon. In Thomas's words, "Dr. Blalock had observed many [times] but had assisted on only one [of these procedures] before he performed his first Blue Baby operation."[29]

Getting the right needles for the delicate vascular operation was a challenge. Standard 1⅛-inch surgical needles were too long to be used in the chests of small children. Thomas's solution was to fabricate smaller needles himself. "Cutting, sharpening, and threading needles," he wrote, "had been part of my routine for all the experimental vascular procedures that I had been performing."[30] Once the surgery had been scheduled, Thomas recalled, "Communication among Dr. Blalock, myself, and the operating-room staff was kept open so that everything was in order."[31] In deciding to operate on Eileen Saxon, the team was entering uncharted territory. The delicate blood vessels in close proximity to the heart, the age and size of the patient, and her poor health all carried grave risks.

The morning of November 29 arrived. Thomas had already told Clara Belle Puryear, a chemistry technician in the Hunterian surgery laboratory, that he didn't plan to observe the operation, jokingly saying he was afraid it would make Blalock nervous. As soon as Blalock entered the operating room and before he scrubbed, he looked up and saw Puryear in the gallery and said, "Miss Puryear, I guess you'd better go call Vivien." Thomas arrived and seated himself in the gallery, but Blalock immediately said, "Vivien, you'd better come down here." According to Thomas, "Dr. Blalock asked me to stand where I could see what he was doing. The best vantage point was on a step stool placed so that I could look over his right shoulder. Dr. Taussig stood by near the head of the table with the anesthesiologist"[32] (figure 10.4). The entire procedure took about two hours, with Blalock frequently consulting with Thomas, whom he could hear but not see over his right shoulder. Snippets of Blalock's tense, sometimes plaintive operating room exchanges with Thomas, as recalled by Blalock's trainees, reveal just how heavily the surgeon depended on his expert and highly skilled technician:

> BLALOCK: Will the subclavian reach the pulmonary once it's cut off and divided?
> BLALOCK: Is the incision long enough?
> THOMAS: Yes, if not too long.
> BLALOCK: Is this all right, Vivien?
> THOMAS: The other direction, Dr. Blalock.
> BLALOCK: Now you watch, Vivien, and don't let me put these sutures in wrong![33]

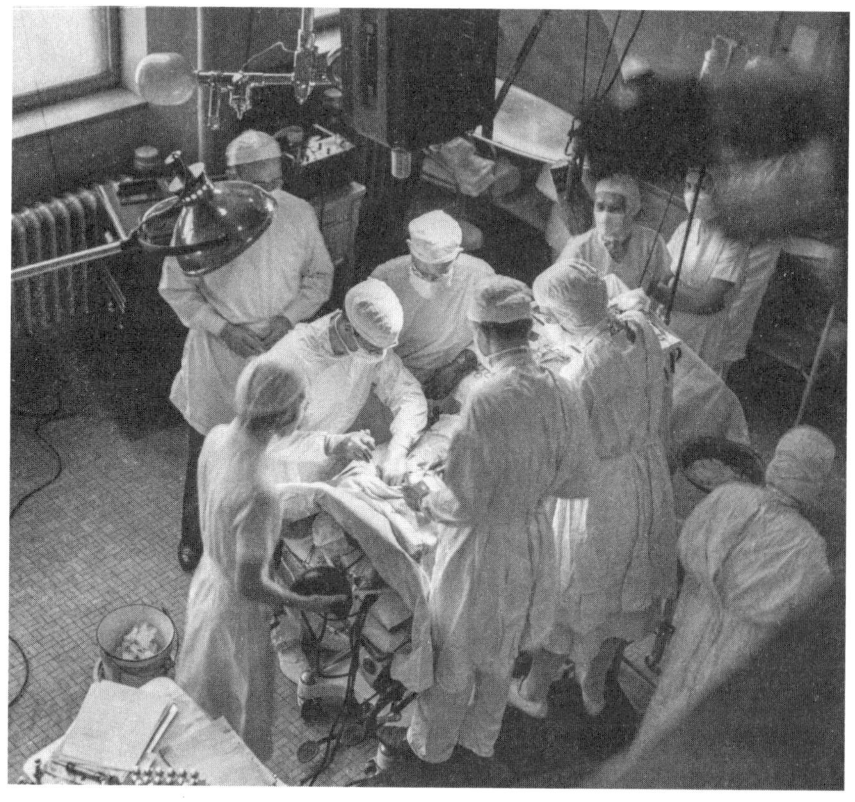

FIGURE 10.4. *Thomas is standing immediately behind Blalock with his fingers just touching in this photograph of a blue baby operation taken in 1947.*

And most telling of all, whenever a member of the surgical team inadvertently stepped into Vivien's appointed space behind Blalock, the surgeon would sternly admonish the offender, saying, "Only Vivien is to stand there."[34]

After the tiny Eileen Saxon's initial postoperative improvement, her recovery slowed. She remained in pediatric intensive care for about two weeks. After two more months, she was released from Hopkins in better condition and with her complexion almost pink. Sadly, as we noted in chapter 9, she died of cardiac complications before turning two. In February 1945, Blalock and his team performed the blue baby procedure on two more children, ages six and twelve, with even better results. Thomas had performed about two hundred operations on dogs to perfect the surgical technique for the subclavian-to-pulmonary artery connection that restored oxygenated blood to the bodies of children afflicted with the congenital tetralogy heart defect. Despite the feeling of elation on the part of the team that had conceived and carried out the operation, it quickly became apparent that Thomas's work had only begun.

Before the pioneering surgery could be turned into a routine procedure, new tools had to be created. Thomas had already produced new needles for the first blue baby operation, but few if any additional instruments for vascular surgery were commercially available. Thomas either designed or adapted from existing implements many of the delicate instruments used in subsequent blue baby operations, including the long bulldog clamp, today known as the Blalock clamp, used to stabilize the major arteries while they were being joined (figure 10.5). After Thomas designed them, Blalock would use his contacts and influence with local surgical supply houses to get these new tools manufactured.

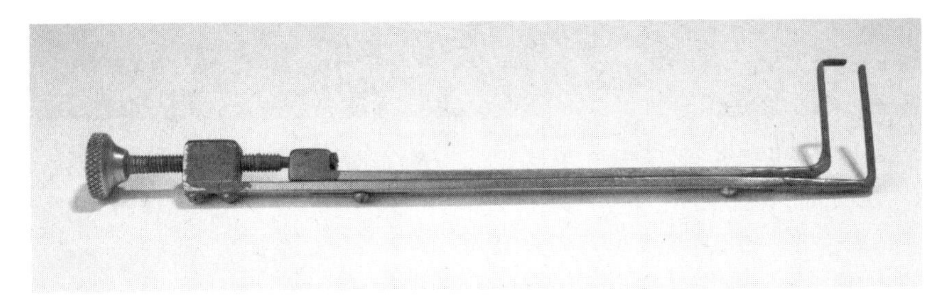

FIGURE 10.5. *Thomas designed and made several instruments, such as this surgical clamp, specifically for the blue baby operation.*

Hopkins at the Center of the Storm

The demands on Thomas's time multiplied when Blalock and Taussig published their historic article "The Surgical Treatment of Malformations of the Heart" in the *Journal of the American Medical Association* (*JAMA*) in May 1945. (Vivien Thomas's name appeared nowhere in its pages.) An Associated Press reporter got hold of the story even before physicians received their copies of the publication in the mail, and the word was out. "Needless to say," Thomas recalled, "Johns Hopkins was not the same for a long time thereafter."[35]

The floodgates were open, and Baltimore became a magnet for parents who had read or heard about the lifesaving blue baby operation. In Thomas's words, "They came by automobile, train, and plane. Many had not communicated with the hospital, had no appointment in the clinic, and had no hotel reservations; thus the cardiac clinic was overrun with patients. At first they were from all over the United States, but after a few weeks, they began arriving from abroad."[36] Adding to the maelstrom of fevered activity swirling around Hopkins were the doctors who came from across the United States and soon after from many foreign countries to watch Blalock and his surgical team operate.

Thomas pushed himself to the brink of exhaustion in those first weeks and months after the publication of the *JAMA* article to keep up with his frenzied workload. Blalock had come to count on Thomas so much that the surgeon routinely demanded his presence in the operating room. In 1945, Blalock and his team completed over one hundred blue baby operations in room 706 of the dispensary building, also known as the heart room. Thomas estimated that, despite their increasingly routine nature, Blalock insisted he stand by his side for at least the first fifty.[37]

Being by Blalock's side in the operating room was only one of Thomas's responsibilities. He would race from the operating room to the wards to draw blood and determine the oxygen levels and blood chemistry of the next group of blue babies scheduled to undergo the procedure. His days became even more hectic when Blalock added postoperative blood studies to Thomas's duties. Eventually, there were additional patients to deal with, patients who had been discharged months earlier and were returning for follow-up blood studies. By then, Thomas's day generally began at 7:30 A.M. with collecting arterial blood samples on the wards, included several round trips to the operating room to support Blalock, and ended well past 8:00 P.M., after he had returned to the surgical laboratory to analyze more blood samples. Somewhere in the middle of the day, Thomas wrote, he stole ten or fifteen minutes to "relax, drink a Coke, and smoke."[38]

Thomas was by nature, and perhaps by necessity, a private man. With his exhausting schedule and his contacts with patients limited to drawing blood, it would be easy to assume that he remained separated from the emotional dramas unfolding around him in Hopkins's operating room and pediatric wards. This simply wasn't the case. In his autobiography, Thomas described one memorable moment:

> A two-and-a-half-year-old girl had never made an attempt to walk or even to stand without coaxing. Her mother broke into tears on entering her room, less than two weeks after the little girl had the Blue Baby operation, to find her standing at the rail of her crib. Scenes like this and seeing the marvelous, almost miraculous improvement in the condition of these little patients . . . had given all of us the strength and stamina to carry on.[39]

Teaching Surgeons by Day, Serving Drinks at Night

Finally comfortable with performing the blue baby surgeries without Thomas by his side, and able to shift the time-consuming blood studies to another department, Blalock, by late 1945, was eager to have Vivien return to his full-time research work. Taussig's challenge to surgically solve the riddle of cyanotic babies was now acting as a catalyst for further research into a whole range of cardiovascular problems. Once he got back to the laboratory, Thomas set to work on several of them, starting with the question of what would happen to the sutures joining the subclavian and pulmonary arteries in infants who underwent the blue baby operation as their bodies grew.

Effectively the manager of Blalock's experimental laboratory, in practice if not in title, Thomas was not only a gifted surgical researcher but also a superb teacher (figure 10.6). In his capacity as the lab's manager and because he had a leading role in a well-regarded animal surgery course, Thomas trained an entire generation of surgical residents, many of whom went on to become leaders in their fields. Denton Cooley, later the founder and surgeon-in-chief of the Texas Heart Institute, was Blalock's surgical intern during the early blue baby operations. "Even if you'd never seen surgery before," Cooley said, "you could do it because Vivien made it look so simple." Of Thomas's surgical technique, Cooley commented, "There wasn't a false move, not a wasted motion, when he operated."[40] Rowena Spencer, a medical student and resident at Hopkins who became a prominent pediatric surgeon, has remarked, "Many times in my career I was complimented on my surgical technique and I will admit that a good many people were shocked when I told them I learned my surgical technique from a black man who had only a high school education."[41]

FIGURE 10.6. *Thomas teaching in 1978.*

It was Blalock himself who paid the ultimate compliment to Thomas's surgical talents. Though Blalock had pioneered a bold and creative solution to the congenital defect that afflicted blue babies, the problem of how to help patients born with the great vessels of the heart transposed had baffled him for years. One day in 1947, Blalock visited the surgical laboratory, and Thomas showed him a solution he had developed. By creating and then suturing a minute hole between the left and right sides of the heart, Thomas had helped address the transposition problem. In his autobiography Thomas wrote:

> Neither he nor I spoke for some four or five minutes while he stood there examining the [dog's] heart, running the tip of his finger back and forth through the moderate-size defect in the atrial septum . . . Dr. Blalock finally broke the silence by asking, "Vivien are you sure you did this?" I answered in the affirmative, and then after a pause, he said, "Well, this looks like something the Lord made."[42]

Despite Thomas's extraordinary skills, his race and status as a mere "technician" often put him in a subservient position. When the Blalocks entertained at their home, Vivien was routinely hired as a waiter or bartender. This sometimes meant that he fixed drinks in the evening for the same residents that he instructed in surgical technique during the day. The disconnect would be glaring today, but William Grose, who trained as a resident under Blalock at Hopkins, recalled a time when "I don't think any of us realized how bizarre it was to have a person of his talents waiting tables or waiting the bar."[43] Blalock apparently had little difficulty separating (and simultaneously juxtaposing) Thomas's roles as a valued member of his professional team and as a household attendant. The surgeon Mark Ravitch, who trained under Blalock, noted that Thomas was not on the guest list in 1960 when a gala was held at Baltimore's elegant—and segregated—Southern Hotel to celebrate Blalock's sixtieth birthday. "The question came up whether Dr. Blalock would want Vivien to be present . . . At that time, David [Sabiston] was the individual who had the warmest personal relationship with Dr. Blalock and it was agreed that he would approach Dr. Blalock on the matter. He returned with the information that Vivien was not to be invited."[44]

Segregation at Hopkins

The Johns Hopkins that Vivien Thomas joined when he followed Blalock to Baltimore in 1941 was a thoroughly segregated institution. Separation of the races and inequality in status, access, and accommodations between whites and "coloreds" prevailed in virtually every area of the institution. In 1992, Louise Cavagnaro, a longtime administrator of Johns Hopkins Hospital, took it upon herself to write a report titled "A History of Segregation and Desegregation at the Johns Hopkins Medical Institution."[45] As an administrator and vice president of the hospital, she played an active role in desegregating patient areas at Johns Hopkins. Cavagnaro updated the document in 2002, but it remains unpublished in the Alan Chesney Medical Archives at Hopkins. The information in this section about segregation and its slow unwinding at Hopkins comes largely from Cavagnaro's typescript report.[46]

Johns Hopkins left instructions in a letter to the trustees that the hospital to be founded in his name shall admit "the indigent sick of this city and its environs, without regard to sex, age, or color."[47] Nonetheless, by 1892, only three years after the hospital opened, the minutes of the trustees noted the desirability of creating "a colored ward" because of the "mutual discomfort" resulting from "the presence of colored patients in the wards with other patients" and the necessity of freeing up space in rooms originally intended for paying patients.[48]

The new building for Black patients opened in March 1894.[49] It housed male and female Black patients until it was remodeled in 1923. Still, separate inpatient units for Black and white adults remained a consistent feature of care at Hopkins until desegregation began with private rooms in the 1950s. The last wards to be desegregated were in psychiatry, where previously Black people requiring inpatient services had always been referred to state hospitals. As with many other policies concerning segregation, the precise time and manner in which they were ended are often difficult to determine.

Fear of mixing the races, and especially their bodily fluids, governed virtually every aspect of care at Hopkins. In the 1940s, the hospital's blood bank had separate refrigerators for "white blood" and "colored blood," with individual bottles labeled accordingly.[50] The hospital superintendent's 1916 annual report stated that facilities for storage of bodies of patients who died in the hospital were currently "unsatisfactory." Accordingly, separate cold rooms—one for Black patients and one for white patients—were constructed in the Pathology Building and remained in use until 1960.

A child's innocent encounter with a water fountain highlights another disturbing feature of segregation at Hopkins. The story is told about a little girl who had taken a drink from a fountain labeled "colored" and ran back to her mother, upset that there were no rainbow hues in the water. Other public facilities were designed to be separate but not always equal. In addition to water fountains, the bathrooms, locker rooms, cafeterias, and operating room waiting areas were segregated. There were two dining areas in the hospital's old Service Building. White employees were welcome in the second floor Nurses Cafeteria, while Black workers, including nurses, technicians, and other professional staff, took their meals in the first floor Employee Cafeteria. In the 1950s, some Black employees began to dine in the Nurses Cafeteria, but the segregated eating arrangements did not completely disappear until the opening of a new cafeteria in 1963.

Vivien Thomas encountered few if any Black professionals in his thirty-eight years at Hopkins—and then only toward the end of his tenure. Discrimination in recruitment, pay, and promotion were facts of life at Hopkins medical institutions from their founding until the latter half of the twentieth century. Black people were most frequently employed in housekeeping and food service positions and to a lesser extent as orderlies. A wage table from 1935 shows different pay scales for Black and white employees.

The first Black student to be admitted to the medical school was Jim Nabwangu, a student from Nigeria. The following year, Robert Gamble, a magna cum laude graduate of Howard University, transferred into the second-year class. Both earned their MDs in 1967. By 2000, 11 percent of medical students at Hopkins were Black.

Fighting Discrimination

As Cavagnaro's report points out, the history of segregation and desegregation at Hopkins is complex and, at crucial points, largely undocumented.

As a hospital, not a governmental agency, many policies that separated the races were not officially recorded; they were more a matter of custom and practice than written regulation. Certainly, when it came to the dismantling of segregation at Hopkins, a trend that gained rapid momentum with the civil rights movement in the early 1960s and federal Affirmative Action policies enacted in its wake, the board and senior leaders at the medical institutions chose to let certain discriminatory measures fade into history, in Cavagnaro's words, "without pronouncements."[51]

Segregation and desegregation at Hopkins were also strongly influenced by "local conditions."[52] As Thomas discovered much to his dismay in moving north from Nashville, Baltimore was very much a southern city. The larger metropolis that lay beyond Hopkins's East Baltimore campus was as thoroughly segregated as the medical institutions were, and vestiges of Jim Crow–era thinking persisted throughout most of Thomas's career. His accomplishments are even more remarkable considering the racial barriers in everything from housing to educating his children that Thomas faced as a Black man in his adopted city.

Later Career and Belated Recognition

With the Supreme Court's ruling in *Brown v. Board of Education* in the 1950s and the emergence of the civil rights movement in the 1960s, racial attitudes at Hopkins and in Baltimore were slowly beginning to change. So, too, was Thomas's situation. In 1952, Thomas built an attractive contemporary-style house in a desirable but previously segregated neighborhood north of Baltimore's baseball stadium. As Thomas noted in his autobiography, the home was in the kind of neighborhood with trees and lawns that he had sought for his family when he moved to Baltimore in 1941 but that had been unavailable to Black people at that time. His productive research partnership with Blalock continued, resulting in many medical journal articles bearing the surgeon's name, though rarely, if ever, was Thomas included as a coauthor. By the late 1950s, Blalock's health had begun to fail. Though Thomas continued to play a critical role in training Blalock's surgical residents, a group that came to be known as the "Old

Hands," the pace of their research slowed considerably. Blalock retired as chief of surgery in 1964, almost at the same time that Hopkins integrated its wards and cafeterias. He died of cancer the following year. There followed a period of nearly six years when Thomas seemed dispirited and did almost no research while still managing the surgical laboratory.

The arrival of the new decade changed things for Thomas. He found meaning and purpose in mentoring a new generation of Black men and women who, like himself, were destined to be pioneers at Hopkins. Richard Scott, the first African American surgical intern at Hopkins, has said of Thomas, "He was my initial supporter. It was as a result of his encouragement that I was able to persevere in some of my career choices."[53]

In 2003 PBS aired an excellent documentary film based on Vivien Thomas's autobiography that shares the title *Partners of the Heart*. Discussing Thomas's renewed sense of himself as a role model for Black medical students, the film notes the generational divide between Thomas and the students he was mentoring. Levi Watkins, the first African American to join the cardiac surgery faculty at Hopkins said:

> We were from different eras. He was [of] the 40s, I was [of] the 50s and 60s. I had been exposed to Dr. King and the whole civil rights movement. I was looking at what was happening at Hopkins, which was not too different from Alabama, to be honest. And so I was getting outspoken about what we should do about black medical students, interns, residents, faculty, so forth. And in a fatherly fashion, he just wanted me to be cool. And wait for the appropriate times.[54]

Thomas had built his career and reputation while being virtually the only African American professional in two deeply segregated hospitals. Of necessity, he picked his battles carefully. Even his use of "Negro" (as opposed to "Black") in his autobiography to describe members of his race reflected his cautious, deliberate approach. Watkins had a more direct, occasionally confrontational, style. The two men respected and learned from each other and together advanced the cause of racial equality at Hopkins (figure 10.7).

FIGURE 10.7. *Four generations in front of the Billings Administration Building at Hopkins. Thomas (right) with the surgical resident Reginald Davis (center), his son James, and the cardiac surgeon and faculty member Levi Watkins (left), 1979.*

By the 1970s, Thomas had spent three decades at Hopkins, and his work with Alfred Blalock had put Johns Hopkins's surgical excellence on the map nationally and internationally. Recognition was long overdue and soon began to come. In February 1971, Thomas, his family, and scores of physicians and staff were present in an auditorium at Hopkins when the portrait that the Old Hands had commissioned and asked him to sit for was unveiled (color plate 14). The painting now hangs opposite Blalock's in the Blalock Clinical Science Building in the main hospital. Thomas said being honored in this fashion was "the most emotional and gratifying experience of my life."[55] At Johns Hopkins University's commencement ceremonies on May 21, 1976, Thomas was awarded an honorary Doctor of Laws degree conferred by the president and the board of trustees (figure 10.8). State licensing regulations prohibited Thomas from receiving an honorary MD, but he had finally earned the fitting and proper title of Dr. Thomas. A short time later, Thomas was appointed to the faculty rank of instructor in surgery at Johns Hopkins. He carried out those duties for three years before retiring in 1979.

FIGURE 10.8. *Thomas (far left) received an honorary doctorate from Hopkins in 1976. Also shown are Helen Taussig (second from left), university president Steven Muller (center), as well as (on the right) Drs. Lubov Keefer and Lewis Thomas.*

Thomas died of pancreatic cancer on November 26, 1985, days before his autobiography, *Partners of the Heart: Vivien Thomas and His Work with Alfred Blalock*, was published. The book inspired both the Emmy-winning HBO film *Something the Lord Made* and the PBS documentary *Partners of the Heart*.

Thomas in His Own Words

In the preface to his memoir, Thomas explained his motivation for putting his life story down on paper. He wrote, "During my career I gave little thought to writing my autobiography or memoirs. However, as the years roll by and my life and my earlier activities became more and more public, I inevitably found myself thinking over the record of years past."[56] Thomas noted that he had become "associated with many men who accomplished so much" in surgical research. Despite lacking "the formal education and training needed to become an instructor in surgery or an administrator of a research facility . . . I served in both capacities."[57] He clearly felt the need to set the record straight. Thomas's 232-page memoir, which relies extensively on his personal files and laboratory notes, does just that, often with the unique insight into the pressures and personalities involved in high-stakes research that only an insider can provide.

Even with this explanation of his motives for writing an autobiography, some have questioned whether a high school–educated Black man could have authored this account of a five-decade-long career that played a pivotal role in creating the field of modern cardiac surgery. Thomas is not here to defend his authorship, but fortunately for those of us concerned with getting his life story right, there is compelling evidence that his autobiography is authentic.

The most convincing proof that Thomas wrote his autobiography lies right in front of our eyes, literally in black and white. Box 17 of the Vivien Thomas Collection, in the Alan Mason Chesney Archives at Johns Hopkins, contains the original manuscript of many of the chapters in Thomas's autobiography. If we compare the opening lines of chapter 15 of Thomas's manuscript, handwritten in his neat, even cursive (figure 10.9), the words match almost exactly those found on page 174 of the published version of *Partners of the Heart.* The first sentence of chapter 15 corresponds to Thomas's manuscript word for word.

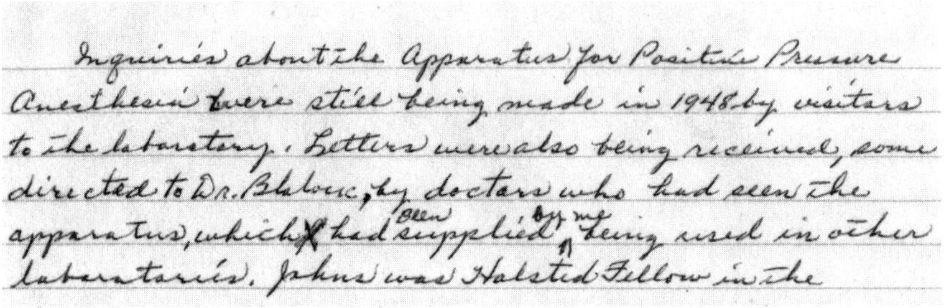

Inquiries about the Apparatus for Positive Pressure
Anesthesia were still being made in 1948 by visitors
to the laboratory. Letters were also being received, some
directed to Dr. Blalock, by doctors who had seen the
apparatus, which ~~I~~ had been supplied by me being used in other
laboratories. Johns was Halsted Fellow in the

FIGURE 10.9. *Thomas's handwritten draft of the opening of chapter 15 of his autobiography. The text in Thomas's neat cursive matches almost word for word page 174 of the same chapter in his published autobiography. The published version starts, "Inquiries about the apparatus for positive-pressure anesthesia were still being made in 1948 by visitors to the laboratory." (Box 17, folder 2 of the Vivien Thomas Collection of the Alan Mason Chesney Archives of the Johns Hopkins Medical Institutions.)*

For every alleged literary fake, there is a supposed forger. Those who won't credit Vivien Thomas with writing his own autobiography routinely point the finger at the surgeon Mark Ravitch, who contributed the book's foreword. Ravitch was Blalock's chief surgical resident in the early 1940s and worked closely with Thomas in the old Hunterian Laboratory. In his foreword to *Partners of the Heart*, Ravitch wrote that it was "a source of satisfaction" that his "prodding" and his office's "secretarial assistance" helped Thomas complete the project.[58] Skeptics have twisted this statement into a claim that Ravitch wrote the book.

Here again, the evidence supports Thomas's authorship. A series of letters exchanged between Ravitch and Thomas between 1980 and 1982 makes it clear that Ravitch's main contributions to Thomas's autobiography were general suggestions about people and topics to include or expand upon. On August 28, 1980, for example, Ravitch wrote Thomas, "I love your comments on the old Hunterian and its horrible condition . . . On page 29, you might tell us a little more about Hel[en] Taussig physically and in manner, and so forth."[59]

Why do some want to question whether Vivien Thomas actually wrote his autobiography? The reasons are not entirely clear, but the effort to cast doubt on his authorship harkens back to the days when physicians and other professionals at Hopkins stared incredulously at a Black man walking the corridors of Hopkins Hospital wearing a white laboratory coat and conducting experimental surgery in the old Hunterian Laboratory. In the authors' view, we do better to celebrate the fact that we have Thomas's vivid, detailed, and readable eyewitness account of the early development of cardiac surgery and his very human struggle for dignity and respect in the face of racism, both overt and subtle.

Although he was not fully credited during his lifetime, Thomas's significant contributions to surgery and surgical science are now fully recognized at Johns Hopkins. One of the four colleges in the medical school is named after Thomas (the others honor Helen Taussig, whom we met in chapter 9; Florence Sabin, a classmate of Dorothy Reed and the first woman elected to the National Academy of Sciences; and Daniel Nathans, winner of the 1978 Nobel Prize in Physiology or Medicine). In May 2021, Johns Hopkins University announced the Vivien Thomas Scholars Initiative. Funded by a $150 million gift from Bloomberg Philanthropies, the Thomas Scholars program addresses historic underrepresentation of minorities in science, technology, engineering, and math disciplines and will train a new, more diverse generation of leaders in these STEM fields.[60] Fittingly named in honor of Thomas, who never had the resources to attend medical school but still went on to do pathbreaking surgical research, the initiative will open new opportunities for students from historically black colleges and universities and minority-serving institutions to pursue doctoral studies at Hopkins.

Recognition and accolades are one thing, but credit where credit is due is quite another. In our view, any fair and equitable assessment of Vivien Thomas's life and contributions to cardiac surgery must acknowledge his pivotal role in developing the blue baby operation. In a 2016 article, the Belgian pathologist J. Van Robays wrote: "To honour his—non-negligible—contribution to the 'Blalock-Taussig shunt,' it should have actually been called the

'Blalock-Thomas-Taussig shunt.' But Vivien Thomas was a person of colour, and in the America of those days, therefore on an even lower social rank than women."[61] Given the scope of his achievements and recognizing the barriers of racism and discrimination Thomas had to overcome, we couldn't agree more.

Epilogue

With Vivien Thomas's quietly determined but ultimately successful struggle to be recognized as an important contributor to the birth of cardiac surgery, we have written the last chapter in our story of the reinvention of American medicine. But like most compelling tales, this account of a modern scientific revolution is hardly at an end. The ten men and women profiled here set in motion a self-perpetuating engine of discovery that is still advancing rapidly. Their life stories hold important lessons for today.

In fact, in an age when there is so much skepticism and mistrust, these condensed biographies are perhaps more important than ever. In a piece titled "This Is How Truth Dies," *New York Times* columnist David Brooks has written that "great nations thrive by constantly refreshing . . . the stories we tell about ourselves."[1] These shared stories, according to Brooks, give us a sense of "who we are . . . what we find admirable . . . [and] what kind of world we hope to build together."[2] Writing a few days before the Fourth of July, Brooks clearly was concerned with America's story. But we would argue that a profession as ancient and important to society's well-being as medicine has its own storytelling tradition that can help us put the lives we have considered—and perhaps more importantly, our own lives and life stories—in proper perspective.

Of course, none of the ten men and women profiled in this book was perfect, and some were deeply flawed. Nevertheless, and even though medicine has moved well beyond their groundbreaking contributions, we believe the arcs of their lives can, to use Brooks's language, "give us a sense of identity, a sense of ideals to live up to and an appreciation of the values that matter most to us."[3] This renewed focus on our "better angels" is a powerful antidote to the cynicism and tribalism that afflict our times.

Each of the ten figures we have chosen has left a unique legacy that has become permanently embedded in the shared story of modern medicine. So, for example, we will always recall Mary Elizabeth Garrett, who had been denied a college education, for the philanthropic jujitsu she performed on Hopkins president Daniel Gilman and his fellow trustees to compel them to admit women to the School of Medicine on an equal footing with men. By the same token, we remember Jesse Lazear, who, with a soldier's courage, sacrificed his life to help prove that the *Aedes aegypti* mosquito is the vector of transmission in yellow fever. And moving closer to our times, we cannot forget Helen Taussig. Her hearing was so impaired that she had to "listen" to the hearts of her young patients with her fingertips, but she dared to imagine the blue baby operation (or, as we have argued, the Blalock-Thomas-Taussig operation), which saved thousands of children born with a congenital heart defect.

We honor these three exemplars—and indeed all ten of our pivotal figures—for their extraordinary human qualities. Their lives demonstrate that, despite its failings and limitations, medicine holds the promise of a better future. Our ten men and women pursued different paths and were possessed of different personalities. Many faced profound adversity. They each combined inner strength with a shared passion. And that passion was a commitment to scientific medicine—medicine that was rooted in research and experimentation, careful clinical observation, and rigorous analysis of data and results. In short, each of them believed that progress in human healing depended on "interrogating" nature. As William Osler, the best-known figure in this extraordinary group, aptly put it, "To have lived through a revolution, to have seen a new birth of science, a new dispensation of health, reorganized medical schools, remodeled hospitals, a new outlook for humanity, is not given to every generation."[4]

We live in an age when even the knowledge we acquire using carefully compiled evidence and rigorous analysis has become a casualty of the all-pervasive culture wars. As Brooks writes, "the collapse of trust" and "the rise of animosity" have become endemic.[5] This state of affairs applies even to knowledge acquired through science and, as our nation's experience with the global COVID-19 pandemic demonstrates, especially through medical research. We don't need to rehash bitter battles over topics like vaccination, masking, and how to keep our children safe in school to know that the belief in research and discovery that these ten men and women worked so hard to create is under attack.

The answer to this corrosive lack of trust in scientific medicine is not easy to discern. In his column, Brooks points out that "these are emotional, not intellectual problems. The real problem is in our system of producing shared stories."[6] As authors, we take some comfort in Brooks's perceptive observation that what we need most are "stories rooted in the complexity of real life" and "narratives in which everybody finds an honorable place."[7]

We have done our level best to tell the stories of ten men and women who made a scientific revolution in American medicine. We have tried to present them in all their human complexity and with an understanding of and empathy for the barriers and constraints they faced as well as the remarkable work they accomplished. It is our hope that sharing their stories will, in some measure, bring us closer together as a community. We leave it to our readers to judge to what degree we have succeeded.

Acknowledgments

We are deeply indebted to Dr. John Cameron, Alfred Blalock Distinguished Service Professor of Surgery, for his inspirational talks on John Shaw Billings and William Stewart Halsted. His love of history is contagious.

We also are truly appreciative of the time and talents of the staff at the Alan Mason Chesney Medical Archives of the Johns Hopkins Medical Institutions. The assistance of Nancy McCall, Andrew Harrison, Timothy Wisniewski, and Phoebe Evans Letocha in making available many of the documents and images in this book is deeply appreciated. So too, are the archival services provided by Tal Nadan, reference archivist at the New York Public Library, who helped us secure important material on John Shaw Billings. Christine Ruggere of the Johns Hopkins Department of the History of Medicine was instrumental in obtaining source material from the department's collections. Kate Long, research services archivist for the Smith College Special Collections; Lily Szczygiel, document technician for the Osler Library of the History of Medicine at McGill University; Rebecca Williams, the archives librarian for research, outreach, and education for the Duke University Medical Center Library and Archives; Emily Bowden, historical collections librarian for the Claude Moore Health Sciences Library of the University of Virginia; Allison Mills, the college archivist for Bryn Mawr College Special Collections; and Claire Hruban, the archivist for the Bryn Mawr School, provided critical assistance in obtaining images and other materials.

Norman Barker, professor of pathology and art as applied to medicine, used his keen eye and deep technical skills to make the historic images in our book as crisp and clean as possible. David Rini, professor in the

Department of Art as Applied to Medicine, artfully blended his skills in traditional fine art techniques with new technologies to create the original illustrations of each of our ten men and women. Rini and Corinne Sandone, the director of the Department of Art as Applied to Medicine at Hopkins, provided thoughtful commentary on Brödel's medical illustrations.

Harper Goldstein provided extensive help as an editorial assistant to this project. We deeply appreciate the many unsung hours of hard work he put in.

We are indebted to the content experts who reviewed selected chapters. Stephen Achuff reviewed the chapter on William Osler, Peter Dawson an early version of the chapter on Dorothy Reed, John Cameron reviewed the chapters on John Shaw Billings and William Stewart Halsted, John Wood and Malcolm Brock the chapter on Vivien Thomas, and Eileen ("Patti") Vining reviewed the chapters on Mary Elizabeth Garrett, Dorothy Reed Mendenhall, and Helen Taussig. Each of them provided us with new perspectives and valuable suggestions. Besides helping us secure dozens of images and documents from the Chesney Medical Archives, Nancy McCall brought her deep knowledge of the history of Hopkins medicine to bear on our project. Claire Hruban read the entire manuscript with her fine critical eye and provided important feedback. Jan Bowers reviewed the chapters as they were being developed and was a key sounding board. Jon Goldstein at Johns Hopkins moderated the online lectures and discussions that gave rise to this book and provided invaluable advice and support throughout the project. We could not have done it without him.

We also would like to recognize Claiborne Hancock and Jessica Case of Pegasus Books, as well as Keith Parent of Simon & Schuster. They gently and thoughtfully shepherded us through the publishing process, making this book a reality.

And finally, a special note of thanks and gratitude to J. Mario Molina for his dedication to celebrating medical history and especially for supporting this book. In the spirit of that philosopher of medicine William Osler, Molina is a true exemplar of what it means to be a physician.

Figure Credits

Alan Mason Chesney Archives of the Johns Hopkins Medical Institutions: Color plates 1, 2, 3, 4, 6, 7, 9, 13, and 14, and figures 1.5, 2.3, 2.4, 2.5, 2.6, 3.1, 3.2, 3.3, 3.4, 3.5, 3.6, 3.7, 4.2, 4.3, 4.4, 5.1, 5.3, 5.4, 5.5, 5.6, 5.7, 6.2, 6.3, 6.4, 6.5, 6.8, 7.1, 7.2, 7.3, 7.9, 8.2, 9.1, 9.2, 9.3, 9.4, 9.5, 9.6, 9.7, 10.2, 10.3, 10.4, 10.5, 10.6, 10.7, 10.8, and 10.9

Baltimore Sun: Figure 5.8

Bryn Mawr School Archives: Figure 1.3

Bryn Mawr College Special Collections: Figures 1.2 and 1.4

David A. Rini, MFA, CMI, FAMI, Professor of Art as Applied to Medicine, Johns Hopkins University: Original portraits at the beginning of each chapter (copyright Department of Art as Applied to Medicine, the Johns Hopkins University).

David Iliff: figure 2.9 (license: CC BY-SA 3.0)

Duke University Medical Center Archives: Figure 10.1

Evergreen House Foundation: Figure 1.1

Johns Hopkins University, Institute of the History of Medicine: Color plate 12 and figure 8.5

Max Brödel Collection, Department of Art as Applied to Medicine, the Johns Hopkins University: Cover Image, figures 7.4, 7.5, 7.6, 7.7, 7.8, and 7.10 and color plate 11

National Library of Medicine: Figures 2.1, 2.2, 4.5, and 4.6

New York Public Library: Figure 2.8

Norman Barker: Figure 2.7

Osler Library of the History of Medicine, McGill University: Figures 4.1 and 4.7

Philadelphia Museum of Art: Color plate 8 (Gift of the Alumni Association to Jefferson Medical College in 1878 and purchased by the Pennsylvania Academy of the Fine Arts and the Philadelphia Museum of Art in 2007 with the generous support of more than 3,600 donors, 2007, 2007-1-1)

Philip S. Hench Walter Reed Yellow Fever Collection, 1806–1995, Historical Collections and Services, Claude Moore Health Sciences Library, University of Virginia: Figures 6.1 and 6.7

Ralph H. Hruban: Figures 1.6 and 6.6

Smith College Special Collections: Figures 8.1, 8.3, 8.4, 8.6, and 8.7

Time magazine: Color plate 5

Washington National Cathedral: Color plate 10

Yale University Library (Yale Athletics Photographs [RU 691], Manuscripts and Archives): Figure 5.2

Bibliography

Abbott, Maude E. "The Pathological Collections of the Late Sir William Osler." *Canadian Medical Association Journal* 10 (1920): 105.

Anderson, Stuart. "Immigrant Nobel Prize Winners Keep Leading the Way for America." Published electronically October 14, 2019. https://www.forbes.com/sites/stuartanderson/2019/10/14/immigrant-nobel-prize-winners-keep-leading-the-way-for-america.

Baker, T. H. "Yellowjack: The Yellow Fever Epidemic of 1878 in Memphis, Tennessee." *Bulletin of the History of Medicine* 42, no. 3 (May–June 1968): 241–264.

Baldwin, Joyce. To Heal the Heart of a Child: Helen Taussig, M.D. New York: Walker, 1992.

Barker, Norman, Ralph Hruban, Alan Wu, and Jon Christofersen. *Halsted: A Documentary* [film]. Johns Hopkins University, 2011.

Barry, John M. *The Great Influenza: The Epic Story of the Deadliest Plague in History*. New York: Viking, 2004.

Billings, John S. *Medical Reminiscences of the Civil War*. Philadelphia, 1905.

———. "Personal Letter to Wife." New York Public Library, 1863.

Blake, John Ballard, and National Library of Medicine (US). *Centenary of Index Medicus, 1879–1979*. Bethesda, MD: National Institutes of Health, 1980.

Blalock, A., and H. B. Taussig. "Landmark Article May 19, 1945: The Surgical Treatment of Malformations of the Heart in Which There Is Pulmonary Stenosis or Pulmonary Atresia. By Alfred Blalock and Helen B. Taussig." *JAMA* 251, no. 16 (April 1984): 2123–2138.

Bliss, Michael. *William Osler: A Life in Medicine*. Oxford: Oxford University Press, 1999.

Brödel, Max. "How May Our Present Methods of Medical Illustration Be Improved?" *JAMA* 49, no. 2 (1907): 138–40.

———. "Medical Illustration." *JAMA* 117, no. 9 (August 30, 1941): 668–672.

———. "Notes on Beer-Making Recipes and Technique." Max Brödel Archives in the Department of Art as Applied to Medicine, Johns Hopkins University.

———. "The Origin, Growth, and Future of Medical Illustration at the Johns Hopkins Hospital and Medical School." *Johns Hopkins Hospital Bulletin* 26, no. 291 (1915): 185–190.

Brooks, David. "This Is How Truth Dies." *New York Times*, July 1, 2021, A21.

Bryan, Charles S. *Osler: Inspirations from a Great Physician*. New York: Oxford University Press, 1997.

———. "Sir William Osler, Eugenics, Racism, and the *Komagata Maru* Incident." *Baylor University Medical Center Proceedings* 34, no. 1 (2021): 194–198.

———. *Sir William Osler: An Encyclopedia*. Novato, CA: Norman Publishing in association with the American Osler Society, 2020.

Bryant, J. E., E. C. Holmes, and A. D. Barrett. "Out of Africa: A Molecular Perspective on the Introduction of Yellow Fever Virus into the Americas." *PLoS Pathogens* 3, no. 5 (May 2007): E75.

Cameron, J. L. "Early Contributions to the Johns Hopkins Hospital by the 'Other' Surgeon: John Shaw Billings." *Annals of Surgery* 234, no. 3 (Sep 2001): 267–78.

———. "William Stewart Halsted. Our Surgical Heritage." *Annals of Surgery* 225, no. 5 (May 1997): 445–458.

Cassedy, James H. *John Shaw Billings: Science and Medicine in the Gilded Age*. Bethesda, MD: Xlibris, 2009.

Cavagnaro, Louise. "A History of Segregation and Desegregation at the Johns Hopkins Medical Institution." Chesney Archives, 1992.

Chapman, Carleton B. "The Flexner Report by Abraham Flexner." *Daedalus* 103 (Winter 1974): 105–117.

———. "John Shaw Billings, 1838–1913: Nineteenth Century Giant." *Bulletin of the New York Academy of Medicine* 63, no. 4 (May 1987): 386–409.

———. *Order Out of Chaos: John Shaw Billings and America's Coming of Age*. Boston: Boston Medical Library, 1994.

Chaves-Carballo, E. "Carlos Finlay and Yellow Fever: Triumph over Adversity." *Military Medicine* 170, no. 10 (October 2005): 881–885.

Chesney, Alan M. *The Johns Hopkins Hospital and the Johns Hopkins University School of Medicine: A Chronicle*. 3 vols. Baltimore: Johns Hopkins University Press, 1943.

Clements, A. N., and R. E. Harbach. "History of the Discovery of the Mode of Transmission of Yellow Fever Virus." *Journal of Vector Ecology* 42, no. 2 (December 2017): 208–22.

Cody, John. "Max Brödel and the Wow! Factor." *Journal of Biocommunication* 37, nos. 2–3 (July 2011): 33–35.

Conway, Jill K. *Written by Herself: Autobiographies of American Women: An Anthology*. New York: Vintage Books, 1992.

Crosby, Ranice W., and John Cody. *Max Brödel: The Man Who Put Art into Medicine*. New York: Springer-Verlag, 1991.

Cullen, Thomas S. "Max Brödel, 1870–1941." *Bulletin of the Medical Library Association* 33, no. 1 (1945): 5–29.

Cushing, Harvey. *The Life of Sir William Osler*. 2 vols. London: Oxford University Press, 1940.

———. "William Stewart Halsted, 1852–1922." *Science* 56 (1922): 461–64.

Cutter, L. "Walter Reed, Yellow Fever, and Informed Consent." *Military Medicine* 181, no. 1 (January 2016): 90–91.

Dawson, Peter J. *Dorothy in a Man's World: A Victorian Woman Physician's Trials and Triumphs*. North Charleston, SC: CreateSpace Independent Publishing Platform, 2016.

———. "Whatever Happened to Dorothy Reed?" *Annals of Diagnostic Pathology* 7, no. 3 (June 2003): 195–203.

del Regato, J. A. "Carlos Juan Finlay (1833–1915)." *Journal of Public Health Policy* 22, no. 1 (2001): 98–104.

———. "Jesse William Lazear: The Successful Experimental Transmission of Yellow Fever by the Mosquito." *Medical Heritage* 2, no. 6 (November–December 1986): 443–452.

———. "The Work of the U.S. Army Board in Havana." Chesney Archives.

Durham, Herbert E., and Walter Myers. "Liverpool School of Tropical Medicine: Yellow Fever Expedition." *British Medical Journal* (1900).

Eberle, F. C., H. Mani, and E. S. Jaffe. "Histopathology of Hodgkin's Lymphoma." *Cancer Journal* 15, no. 2 (March–April 2009): 129–137.

Engle, M. A. "Dr. Helen Brooke Taussig, Living Legend in Cardiology." *Clinical Cardiology* 8, no. 6 (Jun 1985): 372–374.

———. "Growth and Development of Pediatric Cardiology: A Personal Odyssey." *Transactions of the American Clinical and Climatological Assocation* 116 (2005): 1–12.

Farrell, Joseph Pierce, and Peter Occhiogrosso. *Manifesting Michelangelo: The Story of a Modern-Day Miracle—That May Make All Change Possible.* New York: Atria Books, 2011.

Fleming, Donald. *William H. Welch and the Rise of Modern Medicine.* Boston: Little, Brown, 1954.

Flexner, Abraham, and Henry S. Pritchett. *Medical Education in the United States and Canada: A Report to the Carnegie Foundation for the Advancement of Teaching.* New York: Carnegie Foundation for the Advancement of Teaching, 1910.

Flexner, Simon, and James Thomas Flexner. *William Henry Welch and the Heroic Age of American Medicine.* Baltimore: Johns Hopkins University Press, 1993.

Foster, K. R., M. F. Jenkins, and A. C. Toogood. "The Philadelphia Yellow Fever Epidemic of 1793." *Scientific American* 279, no. 2 (August 1998): 88–93.

Garrison, Fielding H. *John Shaw Billings: A Memoir.* New York: G. P. Putnam's Sons, 1915.

Goetsch, Emil. "The Most Unforgettable Character I've Met." *Reader's Digest*, June 1958, 127–34.

Goodman, Gerri Lynn. "A Gentle Heart: The Life of Helen Taussig." Yale Medicine Digital Thesis Library, 1983.

Griffith, Thomas J. "High Points in the Life of Dr. John Shaw Billings." *Indiana Magazine of History* 30 (1934): 325–330.

Haller, J. Alex, Jr. "Anna, Vivien, and the King of Cyan(osis)." *Johns Hopkins Medical Journal* 137 (1975): 55–56.

Halsted, Caroline H. "Caroline Halsted's Letter to Welch." Chesney Archives, 1922.

Halsted, William S. "The Operative Story of Goitre: The Author's Operation." *Johns Hopkins Hospital Report* 19 (1920): 71–257.

———. "Practical Comments on the Use and Abuse of Cocaine." *New York Medical Journal* 42 (1885): 294–295.

———. "The Training of the Surgeon." *Bulletin of the Johns Hopkins Hospital* 15 (1904): 267.

Hamilton, Andrea. *A Vision for Girls: Gender, Education, and the Bryn Mawr School.* Baltimore: Johns Hopkins University Press, 2004.

Hamilton, Edith. "Bryn Mawr School 75th Anniversary." Bryn Mawr School Archives, 1960.

Hartwell, John F. "Letter from Hartwell to Dr. Simon Flexner." Chesney Archives, 1937.

Harvey, A. McGehee. *A Model of Its Kind: Volume 1: A Centennial History of Medicine at Johns Hopkins.* Baltimore: Johns Hopkins University Press, 1989.

———. "John Shaw Billings: Unsung Hero of Medicine at Johns Hopkins." *Maryland Historical Magazine* 84 (1989): 119–134.

Harvey, W. Proctor. "A Conversation with Helen Taussig." *Medical Times* 106, no. 11 (1978): 28–44.

Hinohara, S. "Sir William Osler's Philosophy on Death." *Annals of Internal Medicine* 118, no. 8 (April 15, 1993): 638–642.

"History of 1918 Flu Pandemic." CDC, https://www.cdc.gov/flu/pandemic-resources/1918-commemoration/1918-pandemic-history.htm.

Hogarth, Rana. "A Contemporary Black Perspective on the 1793 Yellow Fever Epidemic in Philadelphia." *American Journal of Public Health* 109, no. 10 (2019): 1337–1338.

Horowitz, Helen Lefkowitz. *The Power and Passion of M. Carey Thomas.* New York: Alfred A. Knopf, 1994.

Hume, Lotta Carswell. *Drama at the Doctor's Gate: The Story of Doctor Edward Hume of Yale-in-China.* New Haven, CT: Yale-in-China Assn., 1961.

Jarrett, W. H. "Raising the Bar: Mary Elizabeth Garrett, M. Carey Thomas, and the Johns Hopkins Medical School." *Baylor University Medical Center Proceedings* 24, no. 1 (January 2011): 21–26.

"Johns Hopkins University School of Medicine Founding Documents." https://jscholarship.library.jhu.edu/handle/1774.2/44451.

Johnson, C. M., C. J. Yeo, and P. J. Maxwell. "The Gross Clinic, the Agnew Clinic, and the Listerian Revolution." *American Surgeon* 77, no. 11 (November 2011): E229–231.

Jones, Absalom, and Richard Allen. "A Narrative of the Proceedings of the Black People during the Late Awful Calamity in Philadelphia in the Year 1793." *American Journal of Public Health* 109 (October 2019): 1336–1337.

Kelly, Howard A. *Walter Reed and Yellow Fever.* 3rd ed. Baltimore, MD: Norman, 1923.

Knight, James A. "William Osler's Call to Ministry and Medicine." *Journal of Medical Humanities and Bioethics* 7, no. 1 (1986): 4–16.

Lazear, J. W. "Letter Fragment from Jesse W. Lazear to Mabel H. Lazear." Historical Collections, Claude Moore Health Sciences Library, University of Virginia, 1900.

Liebowitz, D. "Carlos Finlay, Walter Reed, and the Politics of Imperialism in Early Tropical Medicine." *Pharos of Alpha Omega Alpha Honor Medical Society* 75, no. 1 (Winter 2012): 16–22.

Loechel, W. E. "The History of Medical Illustration." *Bulletin of the Medical Library Association* 48 (April 1960): 168–171.

Longmire, William P., Jr. *Alfred Blalock: His Life and Times.* Published by the author, 1991.

Lown, Bernard. "Black Blood Must Not Contaminate White Folks." Dr. Bernard Lown's blog, September 3, 2011. https://bernardlown.wordpress.com/2011/09/03black-blood-must-not-contaminate-white-folks.

MacCallum, W. G. *A Text-Book of Pathology.* 2nd ed. Philadelphia: W. B. Saunders Company, 1920.

———. *William Stewart Halsted, Surgeon.* Baltimore: Johns Hopkins University Press, 1930.

Macht, David I. "Osler's Perscriptions and Materia Medica." *Transactions of the American Therapeutic Society* 35 (1936): 69–85.

Macnalty, A. S., C. Sherrington, W. W. Francis, C. Singer, L. Horder, and K. J. Franklin. "Sir William Osler Centenary. Some Recollections." *British Medical Journal* 2, no. 4618 (July 1949): 41–48.

Markel, Howard. *An Anatomy of Addiction: Sigmund Freud, William Halsted, and the Miracle Drug, Cocaine.* New York: Pantheon Books, 2011.

McCabe, Katie. "A Legend of the Heart." *Baltimore Magazine*, 1989, 61–65, 102–105.

McKusick, Victor A. "Brödel's Ulnar Palsy: With Unpublished Brödel Sketches." *Bulletin of the History of Medicine* 23 (September–October 1949): 469–479.

McPherson Mary K. "Mary Elizabeth Garrett, 1854–1915. Philanthropist." in *Notable Maryland Women.* Ed. Winified G. Helmes. Cambridge, MD: Tidewater Publishers. 1977, 146–50.

Mehra, A. "Politics of Participation: Walter Reed's Yellow-Fever Experiments." *American Medical Association Journal of Ethics* 11, no. 4 (April 1, 2009): 326–330.

Meisol, Patricia. "The Changing Face of a Strong Woman." *New York Times*, August 18, 2013.

Mencken, H. L. "H. L. Mencken on Brewing a Drinkable Home Brew." *Baltimore Sun*, September 9, 2014. https://www.baltimoresun.com/citypaper/bcp-hl-mencken-on-brewing-a-drinkable-home-brew-20140909-story.html.

———. "Max Brödel." *Baltimore Sun*, October 28, 1941.

———. "Notes on Journalism." *Chicago Tribune*, September 19, 1926.

Mencken, H. L., and Charles A. Fecher. *The Diary of H. L. Mencken.* New York: Knopf, 1989.

Mesquita, E. T., C. V. Souza Jr., and T. R. Ferreira. "Andreas Vesalius 500 Years—a Renaissance That Revolutionized Cardiovascular Knowledge." *Revista Brasileira de Cirurgia Cardiovascular* 30, no. 2 (March–April 2015): 260–265.

Middleton, Natalie. "Spotlight: The Art of Medicine: Max Brödel." *SciArt Magazine*, June 2019. https://www.sciartmagazine.com/spotlight-the-art-of-medicine.html.

Mintz, Morton. "World Warning System on Bad Drugs Is Urged." *Washington Post*, September 18, 1962.

Mukau, Leslie. "Johns Hopkins and the Feminist Legacy: How a Group of Baltimore Women Shaped American Graduate Medical Education." *American Journal of Clinical Medicine* 9 (Fall 2012): 118–127.

Murphy, Jim. *Breakthrough! How Three People Saved "Blue Babies" and Changed Medicine Forever.* Boston: Clarion Books, 2015.

Nesbit, Andrew J. "Time Me, Gentlemen!" *CC2016 Poster Competition*. American College of Surgeons, 2016, 26–30.

"Newspaper Clipping with Note from Surgeon." Chesney Archives, 1972.

Nicholson, John R. "Reminiscences Classmate Yale 1870." Chesney Archives, 1936.

"Partners of the Heart" [transcript]. *American Experience*, February 10, 2003. http://www.shoppbs.pbs.org/wgbh/amex/partners/filmmore/pt.html.

Persaud, Nav, Butts, Heather, and Berger, Phillip, "William Osler: Saint in a 'White Man's Dominion,'" *Canadian Medical Association Journal* 192 (November 9, 2020): 1414–1416.

Peters, C. A. "Henri Amedee Lafleur." *Canadian Medical Association Journal* 62, no. 6 (June 1950): 607–608.

Phelps, Tim. "Exploring Pen and Ink." Phelps Mandala [blog], May 21, 2021. https://phelpsmandala.com/2021/05/21/exploring-pen-and-ink.

Pierce, John R., and Jim Writer. *Yellow Jack: How Yellow Fever Ravaged America and Walter Reed Discovered Its Deadly Secrets*. Hoboken, NJ: J. Wiley, 2005.

Pietila, Antero. *Not in My Neighborhood: How Bigotry Shaped a Great American City*. Chicago: Ivan R. Dee, 2010.

Pubmed Database. https://pubmed.ncbi.nlm.nih.gov.

Ravitch, Mark M. "Letter to Dean Harvey." Chesney Archives, 1986.

———. "Letter to Vivien Thomas." Chesney Archives, 1980.

Reed, Dorothy. "On the Pathological Changes in Hodgkin's Disease, with Especial Reference to Its Relation to Tuberculosis." *Johns Hopkins Hospital Report* 10 (1902): 133–197.

Reed, W., J. Carroll, A. Agramonte, and J. W. Lazear. "The Etiology of Yellow Fever—a Preliminary Note." *Public Health Papers and Reports* 26 (1900): 37–53.

Reed, Walter. "Letter from Walter Reed to Emilie Lawrence Reed." Philip S. Hench Walter Reed Yellow Fever Collection, Claude Moore Health Sciences Library, University of Virginia, 1900.

Sander, Kathleen Waters. *Mary Elizabeth Garrett: Society and Philanthropy in the Gilded Age*. Baltimore: Johns Hopkins University Press, 2008.

———. "A Pleasure to Be Bought." *Johns Hopkins Magazine*, September 2008.

Silverman, B. D. "William Henry Welch (1850–1934): The Road to Johns Hopkins." *Baylor University Medical Center Proceedings* 24, no. 3 (July 2011): 236–42.

Silverman, M. E. "Andreas Vesalius and de humani corporis fabrica." *Clinical Cardiology* 14, no. 3 (March 1991): 276–279.

Silverman, Mark E., T. Jock Murray, and Charles S. Bryan. *The Quotable Osler*. Philadelphia: American College of Physicians, 2002.

Singer, Charles. *A Short History of Anatomy from the Greeks to Harvey*. 2nd ed. New York: Dover Publications, 1957.

"Smith College History and Traditions." Smith College, https://www.smith.edu/topics/history-tradition.

Soylu, E., T. Athanasiou, and O. A. Jarral. "Vivien Theodore Thomas (1910–1985): An African-American Laboratory Technician Who Went on to Become an Innovator in Cardiac Surgery." *Journal of Medical Biography* 25, no. 2 (May 2017): 106–13.

Stevenson, Jeanne H. "Helen Brooke Taussig, 1898– 'The Blue Baby Doctor.'" In *Notable Maryland Wowen*, edited by Winifred G. Helmes, 368–71. Cambridge, MD: Tidewater Publishers, 1977.

Stoner, Joyce Hill. "Andy Warhol and Jamie Wyeth: Interactions." *American Art* 13 (1999): 58–83.

Taussig, H. B., C. H. Kallman, D. Nagel, R. Baumgardner, N. Momberger, and H. Kirk. "Long-Time Observations on the Blalock-Taussig Operation VIII. 20 to 28 Year Follow-up on Patients with a Tetralogy of Fallot." *Johns Hopkins Medical Journal* 137, no. 1 (July 1975): 13–19.

Taussig, Helen. "Address to Western College for Women." Chesney Archives, 1953.

———. "Autobiographical Facts." Chesney Medical Archives, 1973.

———. "Gairdner Foundation Talk." Chesney Archives, 1959.

———. "Letter to Alfred Blalock." Chesney Archives, 1951.

———. "Letter to Danny Nuckols." Chesney Archives, 1972.

———. "Letter to Denton Cooley." Chesney Archives, 1972.

———. "Letter to President Nixon." Chesney Archives, 1970.

———. "Letter to Robert Glaser." Chesney Archives, 1966.

———. "Phi Delta Gamma Talk Notes." Chesney Archives, 1953.

———. "The Thalidomide Syndrome." *Scientific American* 207 (August 1962): 29–35.

Thomas, M. Carrey. "Letter from Thomas to Mary Elizabeth Garrett." Bryn Mawr College Special Collections, 1893.

Thomas, Vivien T. *Partners of the Heart: Vivien Thomas and His Work with Alfred Blalock: An Autobiography*. Philadelphia: University of Pennsylvania Press, 1985.

Thorwald, Jürgen. *The Patients*. Translated by Richard Winston and Clara Winston. New York: Harcourt Brace Jovanovich, 1972.

"The True Story of Thalidomide in the US." US Thalidomide Survivors, https://usthalidomide.org/our-story-thalidomide-babies-us.

Van Robays, J. "Helen B. Taussig (1898–1986)." *Facts, Views, and Vision in ObGyn* 8 (2016): 183–87.

Ventura, H. O. "Giovanni Battista Morgagni and the Foundation of Modern Medicine." *Clinical Cardiology* 23, no. 10 (October 2000): 792–94.

Venugopal, Raghu. "Reading between the Lines: A Glimpse of the Cushing Files and the Life of Sir William Osler." *Osler Library Newsletter*, no. 82 (1996): 1–4.

"Vivien Thomas Scholars Initiative." Johns Hopkins University, Office of the Provost, https://provost.jhu.edu/about/vivien-thomas-scholars-initiative.

Walcott, William Stuart, Jr. "Family Reminiscences Welch and His Father William Wickham Welch." Chesney Archives, 1935.

Welsh, Lilian. *Reminiscences of Thirty Years in Baltimore*. Baltimore: Norman, Remington Co., 1925.

West, J. B. "Galen and the Beginnings of Western Physiology." *American Journal of Physiology: Lung Cellular and Molecular Physiology* 307, no. 2 (July 2014): L121–128.

"William Halsted Obituary." *Baltimore Sun*, September 8, 1922.

Wilson-Pauwels, Linda. *The Development of Academic Programs in Medical Illustration in North America from 1911 to 1991*. Toronto: University of Toronto, 1993.

Winchester, Paul. *The Baltimore & Ohio Railroad: Sketches from the History of the Baltimore and Ohio Railroad*. Baltimore: Maryland County Press Syndicate, 1927.

Winternitz, Milton C. "Re Welch and Students in Pathology School of Medicine." Chesney Archives, 1937.

Zwitter, M., J. R. Cohen, A. Barrett, and E. D. Robinton. "Dorothy Reed and Hodgkin's Disease: A Reflection after a Century." *International Journal of Radiation Oncology, Biology, Physics* 53, no. 2 (June 2002): 366–375.

Notes

PREFACE

1 Bliss, *William Osler*, 1.

INTRODUCTION

1 Mark E. Silverman, Murray, and Bryan, *The Quotable Osler*, 47.
2 Barker et al., *Halsted: A Documentary.*
3 Crosby and Cody, *Max Brödel*, 264.
4 Vivien T. Thomas, *Partners of the Heart*, 122.

CHAPTER 1

1 Chapman, "The Flexner Report by Abraham Flexner," 107.
2 Barry, *The Great Influenza*, 33.
3 Chapman, "The Flexner Report by Abraham Flexner," 107.
4 Barry, *The Great Influenza*, 32.
5 A. McGehee Harvey, *A Model of Its Kind*, 28.
6 Chesney, *The Johns Hopkins Hospital*, 1:255.
7 Jarrett, "Raising the Bar," 21.
8 Sander, *Mary Elizabeth Garrett*, 22.
9 Sander, *Mary Elizabeth Garrett*, 17.
10 Sander, *Mary Elizabeth Garrett*, 18.
11 Sander, *Mary Elizabeth Garrett*, 84.
12 Sander, *Mary Elizabeth Garrett*, 99.
13 Sander, *Mary Elizabeth Garrett*, 78.
14 Horowitz, *The Power and Passion of M. Carey Thomas*, 76.
15 Horowitz, *The Power and Passion of M. Carey Thomas*, 78.
16 Winchester, *The Baltimore & Ohio Railroad*, 43.
17 Mukau, "Johns Hopkins and the Feminist Legacy," 123.
18 Mukau, "Johns Hopkins and the Feminist Legacy," 123.
19 Winchester, *The Baltimore & Ohio Railroad*, 43.
20 Mukau, "Johns Hopkins and the Feminist Legacy," 123.
21 Mukau, "Johns Hopkins and the Feminist Legacy," 124.
22 Sander, *Mary Elizabeth Garrett*, 125.

23 Edith Hamilton, "Bryn Mawr School 75th Anniversary," 2.
24 Andrea Hamilton, *A Vision for Girls*, 23.
25 Sander, *Mary Elizabeth Garrett*, 145.
26 McPherson, *Notable Maryland Women*, 147.
27 Mukau, "Johns Hopkins and the Feminist Legacy," 124.
28 Chesney, *The Johns Hopkins Hospital*, 1: 16.
29 Mukau, "Johns Hopkins and the Feminist Legacy," 125.
30 Sander, *Mary Elizabeth Garrette*, 156.
31 Mukau, "Johns Hopkins and the Feminist Legacy," 125.
32 Sander, *Mary Elizabeth Garrett*, 164.
33 Mukau, "Johns Hopkins and the Feminist Legacy," 125.
34 Sander, *Mary Elizabeth Garrett*, 172.
35 Mukau, "Johns Hopkins and the Feminist Legacy," 119–120.
36 Mukau, "Johns Hopkins and the Feminist Legacy," 120.
37 Mukau, "Johns Hopkins and the Feminist Legacy," 121.
38 Mukau, "Johns Hopkins and the Feminist Legacy," 120–121.
39 Mukau, "Johns Hopkins and the Feminist Legacy," 120.
40 Sander, "A Pleasure to Be Bought."
41 Sander, *Mary Elizabeth Garrett*, 2–3.
42 Chesney, *The Johns Hopkins Hospital*, 1:298–300, 310–312.
43 Chesney, *The Johns Hopkins Hospital*, 1:298.
44 M. Carrey Thomas, "Letter from Thomas to Mary Elizabeth Garrett."
45 Horowitz, *The Power and Passion of M. Carey Thomas*, 237.
46 "Johns Hopkins University School of Medicine Founding Documents."
47 Cushing, *The Life of Sir William Osler*, 1:388.
48 Cushing, *The Life of Sir William Osler*, 1:388.
49 Horowitz, *The Power and Passion of M. Carey Thomas*, 257.
50 Horowitz, *The Power and Passion of M. Carey Thomas*, 429.
51 Horowitz, *The Power and Passion of M. Carey Thomas*, 78.
52 Jarrett, "Raising the Bar," 24.
53 Alan M. Chesney, *The Johns Hopkins Hospital*, 3:36–37.
54 Horowitz, *The Power and Passion of M. Carey Thomas*, 408.

CHAPTER 2

1 Cameron, "Early Contributions to the Johns Hopkins Hospital," 267.
2 Griffith, "High Points in the Life of Dr. John Shaw Billings," 325.
3 Griffith, "High Points in the Life of Dr. John Shaw Billings," 325.
4 A. McGehee Harvey, "John Shaw Billings," 120.
5 Chapman, "John Shaw Billings, 1838–1913," 388.
6 A. McGehee Harvey, "John Shaw Billings," 121.
7 Cameron, "Early Contributions to the Johns Hopkins Hospital," 268.
8 Chapman, "John Shaw Billings," 389.
9 Chapman, *Order Out of Chaos*, 42–43.
10 Cassedy, *John Shaw Billings*, 15.

11 Billings, *Medical Reminiscences of the Civil War*, 118.
12 Garrison, *John Shaw Billings*, 43, 47.
13 Billings, "Personal Letter to Wife."
14 Garrison, *John Shaw Billings*, 64–65.
15 Cameron, "Early Contributions to the Johns Hopkins Hospital," 271.
16 Chapman, *Order Out of Chaos*, 84.
17 Chesney, *The Johns Hopkins Hospital*, 1:13.
18 Chesney, *The Johns Hopkins Hospital*, 1:14.
19 A. McGehee Harvey, "John Shaw Billings," 126.
20 A. McGehee Harvey, "John Shaw Billings," 126.
21 A. McGehee Harvey, "John Shaw Billings," 127.
22 A. McGehee Harvey, "John Shaw Billings," 127.
23 A. McGehee Harvey, "John Shaw Billings," 128.
24 Chesney, *The Johns Hopkins Hospital*, 1:63.
25 Chesney, *The Johns Hopkins Hospital*, 1:64.
26 Chesney, *The Johns Hopkins Hospital*, 1:16.
27 Cameron, "Early Contributions to the Johns Hopkins Hospital," 271–274.
28 Chesney, *The Johns Hopkins Hospital*, 1:248.
29 Chapman, *Order Out of Chaos*, 113.
30 Chesney, *The Johns Hopkins Hospital*, 1:252.
31 Chesney, *The Johns Hopkins Hospital*, 1:244.
32 A. McGehee Harvey, "John Shaw Billings," 130.
33 Simon Flexner and James Thomas Flexner, *William Henry Welch and the Heroic Age of American Medicine*, 92.
34 Cushing, *The Life of Sir William Osler*, 1:297.
35 Chesney, *The Johns Hopkins Hospital*, 1:255.
36 Harvey, *A Model of Its Kind*, 1.
37 Chapman, *Order Out of Chaos*, 153.
38 Cameron, "Early Contributions to the Johns Hopkins Hospital," 269.
39 Chapman, *Order Out of Chaos*, 173.
40 Blake and National Library of Medicine, *Centenary of Index Medicus*, 32.
41 Pubmed Database.
42 Chapman, *Order Out of Chaos*, 195.
43 Chapman, *Order Out of Chaos*, 196.
44 Chapman, *Order Out of Chaos*, 208.
45 Chapman, *Order Out of Chaos*, 232.
46 Cameron, "Early Contributions to the Johns Hopkins Hospital," 275.
47 Chapman, *Order Out of Chaos*, 239.
48 Chapman, *Order Out of Chaos*, 240.
49 Cameron, "Early Contributions to the Johns Hopkins Hospital," 275–276.
50 Chapman, *Order Out of Chaos*, 191.
51 Chapman, *Order Out of Chaos*, 338.
52 Cameron, "Early Contributions to the Johns Hopkins Hospital," 278.
53 Chapman, *Order Out of Chaos*, 334.

54 Chesney, *The Johns Hopkins Hospital*, 1:30.

CHAPTER 3

1 Simon Flexner and James Thomas Flexner, *William Henry Welch and the Heroic Age of American Medicine*, 423.
2 Barry, *The Great Influenza*, 82.
3 Barry, *The Great Influenza*, 83.
4 Abraham Flexner and Henry S. Pritchett, *Medical Education in the United States and Canada*.
5 Chapman, "The Flexner Report by Abraham Flexner," 109.
6 Chapman, "The Flexner Report by Abraham Flexner," 109.
7 Walcott, "Family Reminiscences."
8 Simon Flexner and James Thomas Flexner, *William Henry Welch and the Heroic Age of American Medicine*, 47–48.
9 Nicholson, "Reminiscences Classmate Yale 1870."
10 Simon Flexner and James Thomas Flexner, *William Henry Welch and the Heroic Age of American Medicine*, 54.
11 Simon Flexner and James Thomas Flexner, *William Henry Welch and the Heroic Age of American Medicine*, 62.
12 Barry, *The Great Influenza*, 49.
13 Simon Flexner and James Thomas Flexner, *William Henry Welch and the Heroic Age of American Medicine*, 82.
14 Chesney, *The Johns Hopkins Hospital*, 1:85.
15 Simon Flexner and James Thomas Flexner, *William Henry Welch and the Heroic Age of American Medicine*, 112.
16 B. D. Silverman, "William Henry Welch (1850–1934)," 241.
17 Simon Flexner and James Thomas Flexner, *William Henry Welch and the Heroic Age of American Medicine*, 135.
18 Simon Flexner and James Thomas Flexner, *William Henry Welch and the Heroic Age of American Medicine*, 131.
19 Simon Flexner and James Thomas Flexner, *William Henry Welch and the Heroic Age of American Medicine*, 130.
20 Fleming, *William H. Welch and the Rise of Modern Medicine*, 69.
21 Simon Flexner and James Thomas Flexner, *William Henry Welch and the Heroic Age of American Medicine*, 134.
22 Simon Flexner and James Thomas Flexner, *William Henry Welch and the Heroic Age of American Medicine*, 5.
23 Winternitz, "Re Welch and Students in Pathology School of Medicine."
24 Simon Flexner and James Thomas Flexner, *William Henry Welch and the Heroic Age of American Medicine*, 159.
25 Welsh, *Reminiscences of Thirty Years in Baltimore*, 19.
26 Simon Flexner and James Thomas Flexner, *William Henry Welch and the Heroic Age of American Medicine*, 167.
27 Fleming, *William H. Welch and the Rise of Modern Medicine*, 150.

28 Fleming, *William H. Welch and the Rise of Modern Medicine*, 133.
29 Chesney, *The Johns Hopkins Hospital*, 1:93.
30 Simon Flexner and James Thomas Flexner, *William Henry Welch and the Heroic Age of American Medicine*, 346.
31 Barry, *The Great Influenza*, 58.
32 Fleming, *William H. Welch and the Rise of Modern Medicine*, 131–132.
33 Fleming, *William H. Welch and the Rise of Modern Medicine*, 159.
34 Osler, "Johns Hopkins Hospital—Inner History of Early Institution," 100.
35 Chesney, *The Johns Hopkins Hospital*, 2:175.
36 Simon Flexner and James Thomas Flexner, *William Henry Welch and the Heroic Age of American Medicine*, 408.
37 Barry, *The Great Influenza*, 68.
38 Barry, *The Great Influenza*, 68.
39 Chesney, *The Johns Hopkins Hospital*, 3:135.
40 Simon Flexner and James Thomas Flexner, *William Henry Welch and the Heroic Age of American Medicine*, 375.
41 Simon Flexner and James Thomas Flexner, *William Henry Welch and the Heroic Age of American Medicine*, 375.
42 Simon Flexner and James Thomas Flexner, *William Henry Welch and the Heroic Age of American Medicine*, 376.
43 Barry, *The Great Influenza*, 190.
44 Barry, *The Great Influenza*, 191.
45 "History of 1918 Flu Pandemic."
46 Simon Flexner and James Thomas Flexner, *William Henry Welch and the Heroic Age of American Medicine*, 243.
47 Hume, *Drama at the Doctor's Gate*, 31.
48 Simon Flexner and James Thomas Flexner, *William Henry Welch and the Heroic Age of American Medicine*, 442.
49 Simon Flexner and James Thomas Flexner, *William Henry Welch and the Heroic Age of American Medicine*, 442.
50 Simon Flexner and James Thomas Flexner, *William Henry Welch and the Heroic Age of American Medicine*, 173.
51 Simon Flexner and James Thomas Flexner, *William Henry Welch and the Heroic Age of American Medicine*, 252.
52 Simon Flexner and James Thomas Flexner, *William Henry Welch and the Heroic Age of American Medicine*, 254.
53 Simon Flexner and James Thomas Flexner, *William Henry Welch and the Heroic Age of American Medicine*, 445.
54 Simon Flexner and James Thomas Flexner, *William Henry Welch and the Heroic Age of American Medicine*, 253.
55 Simon Flexner and James Thomas Flexner, *William Henry Welch and the Heroic Age of American Medicine*, 250.
56 Simon Flexner and James Thomas Flexner, *William Henry Welch and the Heroic Age of American Medicine*, 250.

57 Simon Flexner and James Thomas Flexner, *William Henry Welch and the Heroic Age of American Medicine*, 455.
58 Simon Flexner and James Thomas Flexner, *William Henry Welch and the Heroic Age of American Medicine*, 60.
59 Simon Flexner and James Thomas Flexner, *William Henry Welch and the Heroic Age of American Medicine*, 393.
60 Fleming, *William H. Welch and the Rise of Modern Medicine*, 160.
61 Hartwell, "Letter from Hartwell to Dr. Simon Flexner."
62 Simon Flexner and James Thomas Flexner, *William Henry Welch and the Heroic Age of American Medicine*, 50.
63 Fleming, *William H. Welch and the Rise of Modern Medicine*, 3–4.
64 Caroline H. Halsted, "Caroline Halsted's Letter to Welch."

CHAPTER 4

1 Cushing, *The Life of Sir William Osler*, 2:350.
2 Bryan, *Osler: Inspirations*, vii.
3 Bliss, *William Osler*, 26.
4 Cushing, *The Life of Sir William Osler*, 1:20.
5 Bliss, *William Osler*, 27.
6 Cushing, *The Life of Sir William Osler*, 1:22.
7 Bliss, *William Osler*, 32.
8 Cushing, *The Life of Sir William Osler*, 1:145.
9 Bliss, *William Osler*, 88.
10 Bliss, *William Osler*, 88.
11 Cushing, *The Life of Sir William Osler*, 1:359.
12 William Osler, *Aequanimitas*, 368.
13 Cushing, *The Life of Sir William Osler*, 1:134.
14 Bryan, *Sir William Osler: An Encyclopedia*, 54.
15 Abbott, "The Pathological Collections of the Late Sir William Osler," 105.
16 Cushing, *The Life of Sir William Osler*, 1:202.
17 Bliss, *William Osler*, 133.
18 Bliss, *William Osler*, 134–135.
19 Ventura, "Giovanni Battista Morgagni and the Foundation of Modern Medicine," 792.
20 Bliss, *William Osler*, 138–139.
21 Bliss, *William Osler*, 140.
22 Cushing, *The Life of Sir William Osler*, 1:253.
23 Cushing, *The Life of Sir William Osler*, 1:253.
24 Bliss, *William Osler*, 141.
25 Cushing, *The Life of Sir William Osler*, 1:246.
26 Bliss, *William Osler*, 146.
27 Cushing, *The Life of Sir William Osler*, 1:268.
28 Cushing, *The Life of Sir William Osler*, 1:258.
29 Bliss, *William Osler*, 147.
30 Bliss, *William Osler*, 147.

31 Bliss, *William Osler*, 148.
32 Bliss, *William Osler*, 151.
33 Cushing, *The Life of Sir William Osler*, 1:297.
34 Bliss, *William Osler*, 164–165.
35 Peters, "Henri Amedee Lafleur," 607.
36 Bryan, *Sir William Osler: An Encyclopedia*, 341.
37 Cushing, *The Life of Sir William Osler*, 1:339.
38 Bliss, *William Osler*, 183.
39 Bliss, *William Osler*, 185.
40 Cushing, *The Life of Sir William Osler*, 1:324.
41 David I. Macht, "Osler's Perscriptions and Materia Medica."
42 Cushing, *The Life of Sir William Osler*, 1:454.
43 Osler, *The Principles and Practice of Medicine.*
44 Bliss, *William Osler*, 186.
45 Bryan, *Sir William Osler: An Encyclopedia*, 443.
46 Macnalty et al., "Sir William Osler Centenary. Some Recollections," 42.
47 Cushing, *The Life of Sir William Osler*, 1:349.
48 Bliss, *William Osler*, 191.
49 Bliss, *William Osler*, 191.
50 William Osler, "The Natural Method of Teaching the Subject of Medicine," 1678.
51 Cushing, *The Life of Sir William Osler*, 1:596.
52 Bryan, *Osler: Inspirations*, 167.
53 Osler, *Aequanimitas*, 3–4.
54 Osler, *Aequanimitas*, 5.
55 Osler, *Aequanimitas*, 356–357.
56 Osler, *Aequanimitas*, 368.
57 Osler, *A Way of Life*, 27–28.
58 Osler, *Aequanimitas*, 367.
59 Osler, *Aequanimitas*, 201–202.
60 Osler, *Aequanimitas*, 369.
61 Osler, *Aequanimitas*, 404–405.
62 Bryan, *Sir William Osler: An Encyclopedia*, 5.
63 Bliss, *William Osler*, xiii.
64 Bryan, *Sir William Osler: An Encyclopedia*, 64.
65 Bliss, *William Osler*, 199.
66 Cushing, *The Life of Sir William Osler*, 1:414–416
67 Cushing, *The Life of Sir William Osler*, 1:470.
68 Bliss, *William Osler*, 314.
69 Osler, "The Lumleian Lectures on Angina Pectoris," 697–702.
70 Bliss, *William Osler*, 370.
71 Cushing, *The Life of Sir William Osler*, 2:577.
72 Osler, *Aequanimitas*, 7–8.
73 Bryan, *Osler: Inspirationsn*, 198.
74 Bryan, *Sir William Osler: An Encyclopedia*, 193–194.

75 Hinohara, "Sir William Osler's Philosophy on Death," 638–642.
76 Hinohara, "Sir William Osler's Philosophy on Death," 640.
77 Cushing, *The Life of Sir William Osler*, 1:453.
78 Knight, "William Osler's Call to Ministry and Medicine," 14.
79 Hinohara, "Sir William Osler's Philosophy on Death," 641.
80 Cushing, *The Life of Sir William Osler*, 2:684.
81 Osler, *Aequanimitas*, 19.
82 Persaud, Butts, and Berger, "William Osler: Saint in a 'White Man's Dominion,'" 1414–1416.
83 Venugopal, "Reading Between the Lines," 2.
84 Persaud, Butts, and Berger, "William Osler: Saint in a 'White Man's Dominion,'" 1414.
85 Persaud, Butts, and Berger, "William Osler: Saint in a 'White Man's Dominion,'" 1415.
86 Charles S. Bryan, "Sir William Osler, Eugenics, Racism, and the *Komagata Maru* Incident," 195.
87 Charles S. Bryan, "Sir William Osler, Eugenics, Racism, and the *Komagata Maru* Incident," 197.

CHAPTER 5

1 Markel, *An Anatomy of Addiction*, 229–230.
2 Johnson, Yeo, and Maxwell, "The Gross Clinic, the Agnew Clinic, and the Listerian Revolution," 229–231.
3 Goetsch, "The Most Unforgettable Character I've Met," 127.
4 William S. Halsted, "The Training of the Surgeon," 267.
5 MacCallum, *William Stewart Halsted, Surgeon*, viii, 35, 46.
6 Cushing, "William Stewart Halsted, 1852–1922," 461.
7 Olch, "William S. Halsted's New York Period," 496.
8 MacCallum, *William Stewart Halsted, Surgeon*, 11.
9 Cameron, "William Stewart Halsted. Our Surgical Heritage," 446.
10 Olch, "William S. Halsted's New York Period," 498.
11 Cameron, "William Stewart Halsted. Our Surgical Heritage," 448.
12 Cameron, "William Stewart Halsted. Our Surgical Heritage," 448.
13 MacCallum, *William Stewart Halsted, Surgeon*, 43.
14 MacCallum, *William Stewart Halsted, Surgeon*, 35.
15 MacCallum, *William Stewart Halsted, Surgeon*, 46.
16 Goetsch, "The Most Unforgettable Character I've Met," 132.
17 Nesbit, "Time Me, Gentlemen!" 29.
18 Markel, *An Anatomy of Addiction*, 90–91.
19 Markel, *An Anatomy of Addiction*, 97.
20 Noyes, "The Ophthalmological Congress in Heidelberg," 417.
21 Markel, *An Anatomy of Addiction*, 100.
22 William S. Halsted, "Practical Comments on the Use and Abuse of Cocaine," 294–295.
23 Cameron, "William Stewart Halsted. Our Surgical Heritage," 449.
24 Markel, *An Anatomy of Addiction*, 142.
25 Cameron, "William Stewart Halsted. Our Surgical Heritage," 451.
26 Cameron, "William Stewart Halsted. Our Surgical Heritage," 451.

27 Cushing, "William Stewart Halsted, 1852–1922," 461.
28 J. L. Cameron, email to authors, June 10, 2021.
29 Goetsch, "The Most Unforgettable Character I've Met," 130.
30 Goetsch, "The Most Unforgettable Character I've Met," 130.
31 Goetsch, "The Most Unforgettable Character I've Met," 132.
32 Cameron, "William Stewart Halsted. Our Surgical Heritage," 445.
33 Cushing, "William Stewart Halsted, 1852–1922," 461.
34 Goetsch, "The Most Unforgettable Character I've Met," 127.
35 MacCallum, *William Stewart Halsted, Surgeon*, 91.
36 Barker et al., *Halsted: A Documentary.*
37 MacCallum, *William Stewart Halsted, Surgeon*, 92.
38 MacCallum, *William Stewart Halsted, Surgeon*, 81.
39 Olch, "William Stewart Halsted: A Lecture," 421.
40 Cameron, "William Stewart Halsted. Our Surgical Heritage," 452.
41 Cameron, "William Stewart Halsted. Our Surgical Heritage," 445.
42 MacCallum, *William Stewart Halsted, Surgeon*, 182.
43 Cushing, "William Stewart Halsted, 1852–1922," 464.
44 Olch, "William Stewart Halsted: A Lecture," 421.
45 MacCallum, *William Stewart Halsted, Surgeon*, 183.
46 Osler, "Johns Hopkins Hospital—Inner History of Early Institution," 100.
47 Markel, *An Anatomy of Addiction*, 208.
48 MacCallum, *William Stewart Halsted, Surgeon*, 117.
49 Cameron, "William Stewart Halsted. Our Surgical Heritage," 455–456.
50 Goetsch, "The Most Unforgettable Character I've Met," 127.
51 William S. Halsted, "The Operative Story of Goitre," 71–257.
52 Osler, "Johns Hopkins Hospital—Inner History of Early Institution," 100.
53 Cameron, "William Stewart Halsted. Our Surgical Heritage," 457–458.
54 Cameron, "William Stewart Halsted. Our Surgical Heritage," 458.
55 Markel, *An Anatomy of Addiction*, 239.
56 Markel, *An Anatomy of Addiction*, 242–243.
57 Markel, *An Anatomy of Addiction*, 243.
58 Markel, *An Anatomy of Addiction*, 244.
59 Markel, *An Anatomy of Addiction*, 244.
60 Barker et al., *Halsted: A Documentary.*
61 "William Halsted Obituary."
62 Markel, *An Anatomy of Addiction*, 248.

CHAPTER 6

1 Pierce and Writer, *Yellow Jack*, 13.
2 Pierce and Writer, *Yellow Jack*, 11.
3 J. E. Bryant, E. C. Holmes, and A. D. Barrett, "Out of Africa," e75.
4 Foster, Jenkins, and Toogood, "The Philadelphia Yellow Fever Epidemic of 1793," 92.
5 Foster, Jenkins, and Toogood, "The Philadelphia Yellow Fever Epidemic of 1793," 89.
6 Foster, Jenkins, and Toogood, "The Philadelphia Yellow Fever Epidemic of 1793," 89–90.

7 Foster, Jenkins, and Toogood, "The Philadelphia Yellow Fever Epidemic of 1793," 90.
8 Foster, Jenkins, and Toogood, "The Philadelphia Yellow Fever Epidemic of 1793," 90.
9 Hogarth, "A Contemporary Black Perspective on the 1793 Yellow Fever Epidemic in Philadelphia," 1338.
10 Jones and Allen, "A Narrative of the Proceedings of the Black People," 1336.
11 Baker, "Yellowjack," 241.
12 Baker, "Yellowjack," 249.
13 Baker, "Yellowjack," 246.
14 Foster, Jenkins, and Toogood, "The Philadelphia Yellow Fever Epidemic of 1793," 91.
15 Foster, Jenkins, and Toogood, "The Philadelphia Yellow Fever Epidemic of 1793," 89.
16 Baker, "Yellowjack," 243.
17 Pierce and Writer, *Yellow Jack*, 19.
18 Pierce and Writer, *Yellow Jack*, 85–86.
19 Pierce and Writer, *Yellow Jack*, 98.
20 Bryan, *Sir William Osler: An Encyclopedia*, 862.
21 Pierce and Writer, *Yellow Jack*, 99.
22 del Regato, "Carlos Juan Finlay (1833–1915)," 98.
23 Chaves-Carballo, "Carlos Finlay and Yellow Fever," 882.
24 Chaves-Carballo, "Carlos Finlay and Yellow Fever," 882.
25 Chaves-Carballo, "Carlos Finlay and Yellow Fever," 882–883.
26 Chaves-Carballo, "Carlos Finlay and Yellow Fever," 882.
27 Liebowitz, "Carlos Finlay, Walter Reed, and the Politics of Imperialism," 19.
28 Liebowitz, "Carlos Finlay, Walter Reed, and the Politics of Imperialism," 19.
29 Pierce and Writer, *Yellow Jack*, 81.
30 Liebowitz, "Carlos Finlay, Walter Reed, and the Politics of Imperialism," 19.
31 del Regato, "Carlos Juan Finlay (1833–1915)," 100.
32 Pierce and Writer, *Yellow Jack*, 121.
33 Liebowitz, "Carlos Finlay, Walter Reed, and the Politics of Imperialism," 21–22.
34 Pierce and Writer, *Yellow Jack*, 103.
35 Clements and Harbach, "History of the Discovery of the Mode of Transmission of Yellow Fever Virus," 209.
36 Pierce and Writer, *Yellow Jack*, 120.
37 Pierce and Writer, *Yellow Jack*, 88.
38 del Regato, "Jesse William Lazear," 7.
39 del Regato, "The Work of the U.S. Army Board in Havana," 5.
40 Clements and Harbach, "History of the Discovery of the Mode of Transmission of Yellow Fever Virus," 213.
41 Pierce and Writer, *Yellow Jack*, 144.
42 Clements and Harbach, "History of the Discovery of the Mode of Transmission of Yellow Fever Virus," 213.
43 Clements and Harbach, "History of the Discovery of the Mode of Transmission of Yellow Fever Virus," 214.
44 Durham and Myers, "Liverpool School of Tropical Medicine," 656.
45 Durham and Myers, "Liverpool School of Tropical Medicine," 656.

46 Clements and Harbach, "History of the Discovery of the Mode of Transmission of Yellow Fever Virus," 216.
47 Chaves-Carballo, "Carlos Finlay and Yellow Fever," 883.
48 Cutter, "Walter Reed, Yellow Fever, and Informed Consent," 90.
49 Pierce and Writer, *Yellow Jack*, 148.
50 del Regato, "Jesse William Lazear," 15.
51 del Regato, "Jesse William Lazear," 15–16.
52 Pierce and Writer, *Yellow Jack*, 150.
53 Pierce and Writer, *Yellow Jack*, 150.
54 del Regato, "Jesse William Lazear," 16.
55 del Regato, "Jesse William Lazear," 16.
56 Pierce and Writer, *Yellow Jack*, 154.
57 Lazear, "Letter Fragment from Jesse W. Lazear to Mabel H. Lazear."
58 del Regato, "The Work of the U.S. Army Board in Havana," 7.
59 Kelly, *Walter Reed and Yellow Fever*, 152.
60 del Regato, "Jesse William Lazear," 17.
61 W. Reed et al., "The Etiology of Yellow Fever," 18.
62 Pierce and Writer, *Yellow Jack*, 160.
63 Pierce and Writer, *Yellow Jack*, 160–161.
64 Pierce and Writer, *Yellow Jack*, 156.
65 Pierce and Writer, *Yellow Jack*, 156.
66 Clements and Harbach, "History of the Discovery of the Mode of Transmission of Yellow Fever Virus," 218.
67 Cutter, "Walter Reed, Yellow Fever, and Informed Consent," 90.
68 Kelly, *Walter Reed and Yellow Fever*, 164.
69 Clements and Harbach, "History of the Discovery of the Mode of Transmission of Yellow Fever Virus," 218.
70 Kelly, *Walter Reed and Yellow Fever*, 165.
71 Kelly, *Walter Reed and Yellow Fever*, 158.
72 Pierce and Writer, *Yellow Jack*, 192.
73 A. Mehra, "Politics of Participation," 328.
74 Pierce and Writer, *Yellow Jack*, 196–197.
75 Clements and Harbach, "History of the Discovery of the Mode of Transmission of Yellow Fever Virus," 219.
76 Kelly, *Walter Reed and Yellow Fever*, 179–180.
77 Walter Reed, "Letter from Walter Reed to Emilie Lawrence Reed."
78 Kelly, *Walter Reed and Yellow Fever*, 159.

CHAPTER 7

1 Cody, "Max Brödel and the Wow! Factor," 33.
2 Cody, "Max Brödel and the Wow! Factor," 33.
3 Charles Singer, *A Short History of Anatomy*, 2.
4 West, "Galen and the Beginnings of Western Physiology."
5 West, "Galen and the Beginnings of Western Physiology."

6 M. E. Silverman, "Andreas Vesalius and De Humani Corporis Fabrica," 278.
7 Mesquita, Souza, and Ferreira, "Andreas Vesalius 500 Years," 262.
8 Loechel, "The History of Medical Illustration," 169.
9 Loechel, "The History of Medical Illustration," 169.
10 Crosby and Cody, *Max Brödel*, 21.
11 Cullen, "Max Brödel, 1870–1941," 5.
12 Middleton, "Spotlight: The Art of Medicine."
13 Crosby and Cody, *Max Brödel*, 7.
14 Brödel, "The Origin, Growth and Future of Medical Illustration," 186.
15 Brödel, "The Origin, Growth and Future of Medical Illustration," 186.
16 Brödel, "The Origin, Growth and Future of Medical Illustration," 186.
17 Cullen, "Max Brödel, 1870–1941," 22.
18 Crosby and Cody, *Max Brödel*, 20.
19 Crosby and Cody, *Max Brödel*, 21.
20 Crosby and Cody, *Max Brödel*, 30.
21 Cullen, "Max Brödel, 1870–1941," 28.
22 Cullen, "Max Brödel, 1870–1941," 28.
23 Crosby and Cody, *Max Brödel*, 50.
24 Crosby and Cody, *Max Brödel*, 33.
25 Cullen, "Max Brödel, 1870–1941," 7.
26 Brödel, "How May Our Present Methods of Medical Illustration Be Imrpoved?" 32.
27 Cullen, "Max Brödel, 1870–1941," 7.
28 David Rini, email to the authors, August 26, 2021.
29 David Rini, email to the authors, August 26, 2021.
30 Corinne Sandone, interview with the authors, March 5, 2021
31 Phelps, "Exploring Pen and Ink."
32 Middleton, "Spotlight: The Art of Medicine."
33 Cullen, "Max Brödel, 1870–1941," 13.
34 Cody, "Max Brödel and the Wow! Factor," 34.
35 Brödel, "How May Our Present Methods of Medical Illustration Be Imrpoved?" 30.
36 Crosby and Cody, *Max Brödel*, 284.
37 Crosby and Cody, *Max Brödel*, 222.
38 McKusick, "Brödel's Ulnar Palsy," 469–470.
39 Crosby and Cody, *Max Brödel*, 80.
40 McKusick, "Brödel's Ulnar Palsy," 476-477.
41 Crosby and Cody, *Max Brödel*, 147.
42 Crosby and Cody, *Max Brödel*, 143.
43 Wilson-Pauwels, "The Development of Academic Programs in Medical Illustration," 58.
44 Wilson-Pauwels, "The Development of Academic Programs in Medical Illustration," 58.
45 Brödel, "The Origin, Growth and Future of Medical Illustration," 188–189.
46 Brödel, "Medical Illustration," 670.
47 Crosby and Cody, *Max Brödel*, 152.
48 Crosby and Cody, *Max Brödel*, 152.
49 Crosby and Cody, *Max Brödel*, 95.

50 Cullen, "Max Brödel, 1870–1941," 9.
51 Crosby and Cody, *Max Brödel*, 232.
52 Crosby and Cody, *Max Brödel*, 272.
53 Cullen, "Max Brödel, 1870–1941," 12.
54 Mencken, "Notes on Journalism."
55 Crosby and Cody, *Max Brödel*, 263.
56 Mencken, "H. L. Mencken on Brewing a Drinkable Home Brew."
57 Mencken, "H. L. Mencken on Brewing a Drinkable Home Brew."
58 Brödel, "Notes on Beer-Making Recipes and Technique."
59 Cullen, "Max Brödel, 1870–1941," 5.
60 Cody, "Max Brödel and the Wow! Factor," 35.
61 Crosby and Cody, *Max Brödel*, 259.
62 Crosby and Cody, *Max Brödel*, 264.
63 Crosby and Cody, *Max Brödel*, 156.
64 Mencken and Fecher, *The Diary of H. L. Mencken*, 196.
65 Crosby and Cody, *Max Brödel*, 178.
66 Crosby and Cody, *Max Brödel*, 188.
67 Crosby and Cody, *Max Brödel*, 63–64.
68 Crosby and Cody, *Max Brödel*, 72.
69 Crosby and Cody, *Max Brödel*, 72.
70 Crosby and Cody, *Max Brödel*, 201, 205.
71 Crosby and Cody, *Max Brödel*, 214.
72 Mencken, "Max Brödel," 12.
73 Crosby and Cody, *Max Brödel*, 264.
74 Cullen, "Max Brödel, 1870–1941," 5.
75 Anderson, "Immigrant Nobel Prize Winners Keep Leading the Way for America."

CHAPTER 8

1 Dorothy Reed, "On the Pathological Changes in Hodgkin's Disease," 141.
2 Dorothy Reed, "On the Pathological Changes in Hodgkin's Disease," 141–142.
3 Dorothy Reed, "On the Pathological Changes in Hodgkin's Disease," 191.
4 Conway, *Written by Herself: Autobiographies of American Women*, 191–192.
5 Conway, *Written by Herself: Autobiographies of American Women*, 192.
6 Conway, *Written by Herself: Autobiographies of American Women*, 174.
7 Dawson, *Dorothy in a Man's World*, 6.
8 Dawson, *Dorothy in a Man's World*, 7.
9 Dawson, *Dorothy in a Man's World*, 8.
10 Dawson, *Dorothy in a Man's World*, 29.
11 Dawson, *Dorothy in a Man's World*, 9.
12 Dawson, *Dorothy in a Man's World*, 11.
13 Conway, *Written by Herself: Autobiographies of American Women*, 174.
14 Conway, *Written by Herself: Autobiographies of American Women*, 174.
15 Conway, *Written by Herself: Autobiographies of American Women*, 176.
16 Conway, *Written by Herself: Autobiographies of American Women*, 175–176.

17 Dawson, *Dorothy in a Man's World*, 19.
18 "Smith College History and Traditions."
19 Conway, *Written by Herself: Autobiographies of American Women*, 176.
20 Dawson, *Dorothy in a Man's World*, 30.
21 Dawson, *Dorothy in a Man's World*, 29.
22 Dawson, *Dorothy in a Man's World*, 31.
23 Dawson, *Dorothy in a Man's World*, 33.
24 Dawson, *Dorothy in a Man's World*, 33.
25 Dawson, *Dorothy in a Man's World*, 34.
26 Conway, *Written by Herself: Autobiographies of American Women*, 178.
27 Conway, *Written by Herself: Autobiographies of American Women*, 179.
28 Dawson, *Dorothy in a Man's World*, 57.
29 Dawson, *Dorothy in a Man's World*, 61.
30 Conway, *Written by Herself: Autobiographies of American Women*, 180.
31 Dawson, *Dorothy in a Man's World*, 65.
32 Conway, *Written by Herself: Autobiographies of American Women*, 180.
33 Conway, *Written by Herself: Autobiographies of American Women*, 180.
34 Conway, *Written by Herself: Autobiographies of American Women*, 181.
35 Conway, *Written by Herself: Autobiographies of American Women*, 181.
36 Conway, *Written by Herself: Autobiographies of American Women*, 183.
37 Conway, *Written by Herself: Autobiographies of American Women*, 183.
38 Conway, *Written by Herself: Autobiographies of American Women*, 184.
39 Conway, *Written by Herself: Autobiographies of American Women*, 185.
40 Conway, *Written by Herself: Autobiographies of American Women*, 185.
41 Conway, *Written by Herself: Autobiographies of American Women*, 185
42 Conway, *Written by Herself: Autobiographies of American Women*, 185.
43 Conway, *Written by Herself: Autobiographies of American Women*, 185–86.
44 Conway, *Written by Herself: Autobiographies of American Women*, 186.
45 Conway, *Written by Herself: Autobiographies of American Women*, 186.
46 Conway, *Written by Herself: Autobiographies of American Women*, 190.
47 Dawson, *Dorothy in a Man's World*, 95.
48 Dawson, *Dorothy in a Man's World*, 95.
49 Dawson, *Dorothy in a Man's World*, 93.
50 Dawson, *Dorothy in a Man's World*, 93.
51 Dawson, *Dorothy in a Man's World*, 95.
52 Dawson, "Whatever Happened to Dorothy Reed?" 196.
53 Dawson, "Whatever Happened to Dorothy Reed?" 197.
54 Welsh, *Reminiscences of Thirty Years in Baltimore*, 19.
55 Dorothy Reed, "On the Pathological Changes in Hodgkin's Disease," 133.
56 Eberle, Mani, and Jaffe, "Histopathology of Hodgkin's Lymphoma," 129.
57 Dorothy Reed, "On the Pathological Changes in Hodgkin's Disease," 147.
58 Dawson, *Dorothy in a Man's World*, 109.
59 Dorothy Reed, "On the Pathological Changes in Hodgkin's Disease," 133–196.
60 Dorothy Reed, "On the Pathological Changes in Hodgkin's Disease," 142–143.

61 Dorothy Reed, "On the Pathological Changes in Hodgkin's Disease," 151–152.
62 Dawson, "Whatever Happened to Dorothy Reed?" 202.
63 MacCallum, *A Text-Book of Pathology*, 854, 857.
64 Dawson, "Whatever Happened to Dorothy Reed?" 202.
65 Conway, *Written by Herself: Autobiographies of American Women*, 193.
66 Conway, *Written by Herself: Autobiographies of American Women*, 193.
67 Dawson, *Dorothy in a Man's World*, 119.
68 M. Zwitter et al., "Dorothy Reed and Hodgkin's Disease," 368.
69 Dawson, *Dorothy in a Man's World*, 120–121.
70 Dawson, *Dorothy in a Man's World*, 121.
71 Zwitter et al., "Dorothy Reed and Hodgkin's Disease," 368.
72 Dawson, *Dorothy in a Man's Worlds*, 123.
73 Dawson, *Dorothy in a Man's World*, 123.
74 Dawson, *Dorothy in a Man's World*, 131.
75 Dawson, *Dorothy in a Man's World*, 131.
76 Conway, *Written by Herself: Autobiographies of American Women*, 194.
77 Conway, *Written by Herself: Autobiographies of American Women*, 194.
78 Dawson, *Dorothy in a Man's World*, 148.
79 Dawson, *Dorothy in a Man's World*, 149.
80 Conway, *Written by Herself: Autobiographies of American Women*, 194.
81 Dawson, *Dorothy in a Man's World*, 153.
82 Dawson, *Dorothy in a Man's World*, 153.
83 Dawson, "Whatever Happened to Dorothy Reed?" 199.
84 Dawson, "Whatever Happened to Dorothy Reed?" 199.
85 Dawson, "Whatever Happened to Dorothy Reed?" 199.
86 Dawson, "Whatever Happened to Dorothy Reed?" 200.
87 Sander, *Mary Elizabeth Garrett*, 265.
88 Sander, *Mary Elizabeth Garrett*, 263.

CHAPTER 9

1 Taussig, "Autobiographical Facts," 4.
2 Engle, "Growth and Development of Pediatric Cardiology," 2.
3 Baldwin, *To Heal the Heart of a Child*, 46.
4 Taussig, "Autobiographical Facts," 3–4.
5 Baldwin, *To Heal the Heart of a Child*, 47.
6 Engle, "Growth and Development of Pediatric Cardiology," 3.
7 Taussig, "Gairdner Foundation Talk."
8 Stevenson, *Notable Maryland Women*, 369.
9 Baldwin, *To Heal the Heart of a Child*, 52.
10 Taussig, "Letter to Denton Cooley."
11 Murphy, *Breakthrough!*, 40.
12 Vivien T. Thomas, *Partners of the Heart*, 81.
13 Thorwald, *The Patients*, 16.
14 Baldwin, *To Heal the Heart of a Child*, 58–59.

15 Murphy, *Breakthrough!*, 71.
16 Murphy, *Breakthrough!*, 72–73.
17 Blalock and Taussig, "Landmark Article May 19, 1945."
18 Robays, "Helen B. Taussig (1898–1986)," 186.
19 Engle, "Growth and Development of Pediatric Cardiology," 1.
20 Taussig, "Letter to Alfred Blalock."
21 Engle, "Dr. Helen Brooke Taussig, Living Legend in Cardiology," 372.
22 Taussig, "Letter to Danny Nuckols."
23 Baldwin, *To Heal the Heart of a Child*, 22.
24 Helen Taussig, "Letter to Robert Glaser."
25 Stevenson, *Notable Maryland Women*, 370.
26 Taussig, "The Thalidomide Syndrome," 29.
27 Mintz, "World Warning System on Bad Drugs Is Urged."
28 W. Proctor Harvey, "A Conversation with Helen Taussig," 42.
29 W. Proctor Harvey, "A Conversation with Helen Taussig," 42.
30 Taussig, "The Thalidomide Syndrome," 29.
31 Taussig, "The Thalidomide Syndrome," 35.
32 "The True Story of Thalidomide in the US."
33 Taussig, "Letter to President Nixon."
34 Goodman, "A Gentle Heart," 82.
35 Taussig et al., "Long-Time Observations on the Blalock-Taussig Operation VIII."
36 "Newspaper Clipping with Note from Surgeon."
37 Baldwin, *To Heal the Heart of a Child*, 102.
38 Taussig, "Address to Western College for Women."
39 Taussig, "Phi Delta Gamma Talk Notes."
40 Baldwin, *To Heal the Heart of a Child*, 98.
41 Baldwin, *To Heal the Heart of a Child*, 100.
42 Baldwin, *To Heal the Heart of a Child*, 102.
43 Baldwin, *To Heal the Heart of a Child*, 105.
44 Meisol, "The Changing Face of a Strong Woman."
45 Baldwin, *To Heal the Heart of a Child*, 104–105.
46 Baldwin, *To Heal the Heart of a Child*, 105.
47 Stoner, "Andy Warhol and Jamie Wyeth," 64.
48 Baldwin, *To Heal the Heart of a Child*, 105.
49 Baldwin, *To Heal the Heart of a Child*, 104.

CHAPTER 10

1 Vivien T. Thomas, *Partners of the Heart*, 5.
2 Vivien T. Thomas, *Partners of the Heart*, 6.
3 Vivien T. Thomas, *Partners of the Heart*, 9.
4 Vivien T. Thomas, *Partners of the Heart*, 9.
5 Vivien T. Thomas, *Partners of the Heart*, 10–11.
6 Vivien T. Thomas, *Partners of the Heart*, 12.
7 Vivien T. Thomas, *Partners of the Heart*, 13.

8 Vivien T. Thomas, *Partners of the Heart*, 126.
9 Vivien T. Thomas, *Partners of the Heart*, 34–35.
10 Vivien T. Thomas, *Partners of the Heart*, 10.
11 Vivien T. Thomas, *Partners of the Heart*, 16.
12 Vivien T. Thomas, *Partners of the Heart*, 16.
13 Vivien T. Thomas, *Partners of the Heart*, 17.
14 Longmire, *Alfred Blalock*, 33.
15 Vivien T. Thomas, *Partners of the Heart*, 38.
16 Vivien T. Thomas, *Partners of the Heart*, 49.
17 Vivien T. Thomas, *Partners of the Heart*, 57.
18 Vivien T. Thomas, *Partners of the Heart*, 57.
19 Pietila, *Not in My Neighborhood*, 80.
20 Vivien T. Thomas, *Partners of the Heart*, 60.
21 Vivien T. Thomas, *Partners of the Heart*, 63–64.
22 Vivien T. Thomas, *Partners of the Heart*, 64.
23 "Partners of the Heart" [transcript].
24 Vivien T. Thomas, *Partners of the Heart*, 81.
25 Vivien T. Thomas, *Partners of the Heart*, 88.
26 Haller, "Anna, Vivien, and the King of Cyan(osis)," 56.
27 Haller, "Anna, Vivien, and the King of Cyan(osis)," 56.
28 Vivien T. Thomas, *Partners of the Heart*, 91.
29 Vivien T. Thomas, *Partners of the Heart*, 99.
30 Vivien T. Thomas, *Partners of the Heart*, 91.
31 Vivien T. Thomas, *Partners of the Heart*, 92.
32 Vivien T. Thomas, *Partners of the Heart*, 92.
33 McCabe, "A Legend of the Heart," 65, 102.
34 McCabe, "A Legend of the Heart," 102.
35 Vivien T. Thomas, *Partners of the Heart*, 97.
36 Vivien T. Thomas, *Partners of the Heart*, 97.
37 Vivien T. Thomas, *Partners of the Heart*, 101–102.
38 Vivien T. Thomas, *Partners of the Heart*, 103.
39 Vivien T. Thomas, *Partners of the Heart*, 104.
40 Soylu, Athanasiou, and Jarral, "Vivien Theodore Thomas (1910–1985)," 112.
41 Farrell and Occhiogrosso, *Manifesting Michelangelo*, 226.
42 Vivien T. Thomas, *Partners of the Heart*, 122.
43 "Partners of the Heart" [transcript].
44 Ravitch, "Letter to Dean Harvey."
45 Cavagnaro, "A History of Segregation and Desegregation at the Johns Hopkins Medical Institutions."
46 Cavagnaro, "A History of Segregation and Desegregation at the Johns Hopkins Medical Institution."
47 Chesney, *The Johns Hopkins Hospital*, 1:14.
48 Cavagnaro, "A History of Segregation and Desegregation at the Johns Hopkins Medical Institution," 4.

49 Chesney, *The Johns Hopkins Hospital*, 2:22.
50 Lown, "Black Blood Must Not Contaminate White Folks."
51 Cavagnaro, "A History of Segregation and Desegregation at the Johns Hopkins Medical Institution," 1.
52 Cavagnaro, "A History of Segregation and Desegregation at the Johns Hopkins Medical Institution," 1.
53 "Partners of the Heart" [transcript].
54 "Partners of the Heart" [transcript].
55 McCabe, "A Legend of the Heart," 104.
56 Vivien T. Thomas, *Partners of the Heart*, xi.
57 Vivien T. Thomas, *Partners of the Heart*, xi.
58 Vivien T. Thomas, *Partners of the Heart*, foreword.
59 Ravitch, "Letter to Vivien Thomas."
60 "Vivien Thomas Scholars Initiative."
61 Van Robays, "Helen B. Taussig (1898–1986)," 186.

EPILOGUE

1 Brooks, "This Is How Truth Dies."
2 Brooks, "This Is How Truth Dies."
3 Brooks, "This Is How Truth Dies."
4 Silverman, Murray, and Bryan, *The Quotable Osler*, 259.
5 Brooks, "This Is How Truth Dies."
6 Brooks, "This Is How Truth Dies."
7 Brooks, "This Is How Truth Dies."

Index

C

Y